Physics on the Fringe

BY THE SAME AUTHOR

Pythagoras' Trousers

The Pearly Gates of Cyberspace

A Field Guide to Hyperbolic Space

Physics on the Fringe

Smoke Rings, Circlons, and Alternative Theories of Everything

Margaret Wertheim

Walker & Company
New York

For information address Walker & Company, 175 Fifth Avenue,
New York, New York 10010.

Published by Walker Publishing Company, Inc., New York
A Division of Bloomsbury Publishing

All papers used by Walker & Company are natural, recyclable products made
from wood grown in well-managed forests. The manufacturing processes
conform to the environmental regulations of the country of origin.

LIBRARY OF CONGRESS CATALOGING-IN-PUBLICATION DATA
HAS BEEN APPLIED FOR.

ISBN: 978-0-8027-1513-5

Visit Walker & Company's Web site at www.walkerbooks.com.

First U.S. edition 2011

1 3 5 7 9 10 8 6 4 2

Typeset by Westchester Book Group
Printed in the U.S.A. by Quad/Graphics, Fairfield, Pennsylvania

For my sister Christine Wertheim,
who showed me how to think for myself.

I dream of a new age of curiosity. We have the technical means for it; the desire is there; the things to be known are infinite; the people who can employ themselves at this task exist. Why do we suffer? From too little, from channels that are too narrow, skimpy, quasi-monopolistic, insufficient. There is no point in adopting a protectionist attitude, to prevent "bad" information from invading and suffocating the "good." Rather, we must multiply the paths and the possibility of comings and goings.

MICHEL FOUCAULT

"THE MASKED PHILOSOPHER"

CONTENTS

Part Three: Sciences of Imaginary Solutions

Chapter Zero

A TRAILER PARK OWNER IMAGINES THE WORLD

WHAT DRIVES A man with no science training to think he can succeed where Albert Einstein and Stephen Hawking have failed? In 1993, Jim Carter sent out to a select group of the nation's scientists a letter announcing the publication of his book *The Other Theory of Physics*, in which he promised a complete alternative description of the universe. "Never before has any theory offered such a comprehensive explanation," his letter enthused. In addition to "shedding new light on phenomena that have long been considered to be well explained," the book would contain solutions to "mysteries and paradoxes that have plagued physical science for centuries." The "origin of the moon," the nature of gravity, the structure of matter, the relationship between space and time—all these Carter proposed to clarify through a concept he called "Circlon Synchronicity." This was not an extension of existing physics, but a wholesale reconstruction. If his ideas were accepted, it would be the greatest upheaval in science since the Copernican revolution—he would be not just the next Einstein, but the Isaac Newton of our time.

The origin of the author was itself something of a mystery. No college or university was mentioned in his announcement, but his letterhead carried the logo of the Absolute Motion Institute, an organization apparently located on Green River Gorge Road in Enumclaw, Washington. The theory had not been published before or ever presented to the public. It was, Carter stated with disarming honesty, the fruit of more than twenty-five years of solitary labor. For those wishing to acquaint themselves with this work, a small yellow order form was included in the package. Interested parties were invited to tick a box and send in a check: $88 was the asking price for a "premier hardcover edition," while a paperback edition, for $44, was promised the following year. Yet it was three small boxes at the bottom of the form that most drew the attention of a careful reader. Here were offered the following options:

> "I'm very interested in this theory, but at this time I am very short of cash. However, the enclosed letter expressing my response to the ideas of Circlon Synchronicity and Absolute Motion qualifies me to receive a free copy of the premier edition of *The Other Theory of Physics* on its publication date."

> "The idea interests me! Please keep me on the Absolute Motion mailing list for the next free mailing."

> "Your Gravity Theory Sucks!"

That final phrase catapulted Carter into a rare class: The man had a sense of humor. That is a quality almost categorically absent from the majority of "outsider theorists" whose packages arrive at the offices of professional physicists with rather more frequency than many scientists are comfortable admitting.

Being at the time short of cash myself, I sent off a letter with my response to the ideas of Circlon Synchronicity as I had dimly understood them from the materials Carter had provided and several weeks later, I received in the mail a prepress edition of his book. A densely packed assortment of theoretical speculations, empirical findings, and graphical play, it was like no other science book I had seen. In addition to its bamboozling ideas, it was filled with diagrams and charts and doodles and equations, all of them clearly by Carter's hand. Whatever you could say about his physics, he had a marvelous visual style. Here was an entire phantasmagoric universe: atoms, stars, and galaxies; the moon, tides, spaceships, and bumblebees. Scattered throughout the text were concepts both astounding and alarming: "negative matter" bodies, "seven dimensions of time," something Carter called "string demons," and an analysis of what might happen to the *Titanic* if the great ship were traveling at half the speed of light.

By Carter's own assessment, the system reached its apogee in the claim that nature's most inexorable force was an illusion. To quote the book's stark denial, "Gravity does not exist." In its place he proposed "constantly expanding matter," a wildly profligate inflation of each and every particle. You, me, the chair you are sitting on, the trees outside your window—everything that exists—according to Carter, we are all constantly expanding. Every minute of every hour of every day. As a consequence, he claimed, the earth itself doubles in size every nineteen minutes. Hold a pencil in your hand and let it go: As Carter's theory tells it, the pencil doesn't *go* anywhere; rather, the earth rises up to meet it. Or as he would later tell me: "Gravity is not the result of things falling down, but of the earth falling up."

In university physics departments there is a term for people like Jim Carter. They are generally known as "cranks," and the trajectories of their packages are typically short—straight from the mailroom into the bin. Secretaries of well-known physicists quickly learn to spot the telltale signs, for such manuscripts will usually announce themselves by obvious deviations from the standards of scientific practice. In all likelihood there will be an abundant use of CAPITAL LETTERS and exclamation points!!! Important sections will be underlined, or **bolded**, or circled, for emphasis. Frequently the author will have seen fit to ease the professor's path toward understanding by writing helpful comments in the margins of the paper or by highlighting critical passages with brightly colored felt-tip pens. All such marks will cause suspicion in the mind of a credentialed physicist, who can generally recognize a fringe theory by sight, without even reading the text. The text itself will almost certainly herald its revolutionary nature in its opening paragraphs, claiming to reinvent if not the whole of physics, as in Carter's unusual case, then at least substantial parts of it. At a minimum, the author will be proposing something radically new and often as not will have harsh words for the twin pillars of twentieth-century physics—relativity and quantum theory. Typically, he or she will be offering an "alternative," "simpler," "more comprehensible" explanation.

The academic science world can be harsh on men like Jim Carter, and I use the male noun here specifically, for virtually all outsider physics theorists are men. While insider physics remains the most male-dominated of the academic sciences, its outsider equivalents are almost always male. When credentialed physicists choose to comment on such theories at all, their remarks tend to be derisive or at best dismissive. Most often, from the outsiders' perspective, there will be something far worse: silence. Jim Carter bears such oversight stoically—so convinced is he that history will eventually side with his ideas—but for many outsider theorists, the academic world's dismissal of their work is infuriating and heartbreaking. In all sincerity, these men believe they have found "the secrets of the universe," to use one of Jim's favorite terms, and they are eager to share their insights with the rest of us. Like insiders, they too want to illuminate the void of ignorance and expand the domain of scientific knowledge for all human beings. To a man, they believe they have discovered profound and simple truths essential to the workings of nature that mainstream physicists have missed.

For the past fifteen years I have been collecting the work of "outsider physicists," as I have come to fondly think of these people, and I now have on my shelves around a hundred such theories. Some of them are entire books, some are lengthy articles. Some are brief sketches for a theory, while others are fully fleshed out. Some have been professionally printed, some are typewritten, and some, which I especially value, are handwritten and in a few cases also hand-illustrated. My collecting has not been systematic or comprehensive in any way. I have pretty much taken what I have stumbled upon or what has come to me via the post or the Internet through no particular solicitation on my part. Over the years, I have been amazed at how much *has* come my way, for I am not a famous physicist. While it is true that I trained as a physicist and originally thought this would be my profession, after university I decided to become a science writer and have been working at that vocation my entire professional life. As

I am not based at a university and work from home, I cannot be all that easy to find, yet over the years a steady stream of unorthodox theories has found its way to my door. I have kept them, chuckled at them, been delighted and exasperated by them. Sometimes, as in Jim's case, I have been enchanted by them. Slowly, as they massed on a shelf in my office, I found myself increasingly intrigued. What did this outpouring of human effort represent?

In 1995, about eighteen months after I received Carter's announcement, I decided to take a trip to meet the man himself. As a science journalist, I thought he might be an interesting subject for an article, and I did not then imagine writing a book. I had been invited to give a lecture at Pacific Lutheran University in Tacoma, Washington, and after the talk I rented a car and drove out to Enumclaw, about thirty miles away. I had no idea what to expect, and whatever I might have expected I think it's safe to say I would have been wrong. Carter lived in a trailer park on a spectacular piece of land nestled in a forest on the lip of the aptly named Green River Gorge. It turned out that he owned the land and had built the trailer park himself. On his front lawn was a collection of huge vintage Chryslers and Cadillacs rusting in picturesque heaps. There was a pet swan that followed him around like a dog. He also had two actual dogs, plus a wife and two young sons, and the swan must have understood something about his affection for his family because every time his wife came near it, it hissed and spat and tried to bite her. Soon after I met Jim the swan disappeared, and a local gang of trailer park dogs was suspected of the crime. I kept going back, and over the years my encounter with Jim has grown into one of the strangest and most challenging relationships of my life. Quietly and inexorably, as I sought to understand what makes him tick, I began to realize that I had to reassess what makes the science of physics itself tick.

As I intersected with Carter's life and work, I began to wonder what exactly is the role of theoretical physics in the life of our society? What functions does it serve? For whom? And how? And why?

One of the functions physics serves that is frequently mentioned by physicists is that this is the science that helps us to make microchips and other instrumental marvels. Thanks to the quantum understanding of matter, we have semi-conductors and lasers. Thanks to James Clerk Maxwell's equations of electromagnetism, and engineers' application of these equations, we have electricity running efficiently to our homes and the miracle of mobile phone communication. Thanks to our understanding of general relativity and the deviations our planet causes in the structure of spacetime, GPS satellite-based tracking systems can tell us our positions on earth to an accuracy of two or three meters, a truly astonishing achievement that is underlying Google Earth and all its ancillary "apps."

Yet instrumental success is not the only goal of science, because when Einstein came up with the general theory of relativity, nobody thought it would have any practical applications and some physicists suspected it was a fantasy. Einstein wasn't trying to make our lives easier when he beavered away on his legendarily difficult equations; he too was trying to discern the secrets of the universe, the "natural laws," that would, in his words, inspire "a form of rapturous amazement."

Einstein's work points us to another role that theoretical physics serves that in some respects has come to be an equally crucial function: In the modern Western world, this is the subject through which we seek to understand the wider cosmological scheme in which we humans are embedded. "Our goal is a complete understanding of the events around us, and of our own existence," Stephen Hawking tells us in *A Brief History of Time*. If we call ours "the age of science," it is not just because science helps us materially, but because science has also become the frame through which we now try to map reality, displacing religion, which used to service this goal. Ever since Copernicus and his contemporaries in the sixteenth century replaced the earth-centered, God-focused vision of the cosmos with a sun-centered view, the officially sanctioned picture of our universe has increasingly been dictated by astronomy and physics. With the advent of general relativity in the early twentieth century, theoretical physics grew to encompass within its equations the entire space of being—the sun, the moon, the planets, and the stars and the whole arena of space and time. Einstein is not the most famous physicist on earth because he discovered the principles behind lasers and the photoelectric effect, both of which have enormous practical applications; we all know his face and name because his equations showed us in some sense the totality of our world. General relativity gives us our universe in sums.

Scientists themselves increasingly point to this *cosmological* function of their work. One of science's aims, it is often now said, is to help us feel "at home in the universe." That is the title of a fascinating book by complexity theory pioneer Stuart Kauffman. Kauffman's book itself proposes a radical theory about the evolution of the universe that is highly controversial, and many scientists dispute his conclusions; but in offering up his ideas, Kauffman puts forward an argument about what science *achieves* that many of his colleagues find appealing. One of the impulses for writing the book, Kauffman says, was to counter the idea that science alienates humans from the rest of the cosmic scheme. On the contrary, he argues, science shows us how we are an intrinsic part of the universal whole. In Kauffman's view, the more we learn through science, the more we are enabled to feel "at home in a universe" whose beauty and complexity enrich and nourish us. While many scientists may disagree with Kauffman's own theory, his felicitous phrase has become something of a rallying cry to champions of science worldwide.

In his book *Unweaving the Rainbow*, British biologist Richard Dawkins voiced a complementary sentiment. One of science's most ardent public champions, Dawkins tells us here that science leads to a "feeling of awed wonder" by delivering "one of the highest experiences of which the human psyche is capable. It is a deep aesthetic passion to rank with the finest that music and poetry can deliver. It is truly one of the things that makes life worth living." The aesthetic passion that science stirs up is particularly acute in the arena of theoretical physics, which Hawking has famously proposed can now replace God as a foundation for meaning in our lives. For Stephen Hawking, Richard Dawkins, Stuart Kauffman, and many other professional scientists, science provides a platform that is both aesthetically and emotionally satisfying, that is *for them* a foundation for psychic well-being.

As someone who spent five years at a university studying physics and mathematics because I thought they were beautiful, my heart leaps with Dawkins's stirring words. Personally I find the equations of physics to be an exquisite kind of music. Yet my encounters with Jim and other physics outsiders have led me to questions for which there are no easy answers and that point to currents in our culture that are not easily dismissed: *Who* gets to feel "at home in the universe" contemporary science describes? *Who* gets to feel stirred by the world picture portrayed by the equations of general relativity and quantum theory? What motivates Jim and other fringe theorizers is precisely that they don't feel at home in this world. On the contrary, they feel alienated by the accounts of reality they read about in science magazines and books. For these people, the equations of theoretical physics seem like not a symphony, but a cacophony. The majestic formal poetry that so thrills the insiders comes across to their ears as gibberish. As far as Jim is concerned, the "spacetime matrix" of general relativity is nonsense.

Along with their academic counterparts, theoretical physics outsiders are also trying to feel at home in a universe governed by scientific laws. They too believe that science can deliver experiences of awe and wonder and beauty. In contrast with creationists, who also reject much of academic science, fringe physicists love science and are thrilled by its power. Their belief is indeed that, as Francis Bacon promised at the start of the seventeenth century, science can deliver a reasonable, functional, and satisfying description of reality. Their *claim* is that such a description ought to be comprehensible to all human beings.

At the start of the scientific revolution, Bacon urged his contemporaries to stop relying on received wisdom and go out and engage with nature for themselves. That is what Jim Carter is doing. Where Bacon rejected the books of academic scholasticism, so outsider theorists today reject the books of academic physics, insisting instead that they can discern nature's laws through their own experience. Champions of the scientific mainstream may discount Jim's results, but his *efforts* adhere to many of the principles on which modern science was built. In both Bacon's case and Jim's, what is being challenged is a class of authorities who communicate in a language that often seems remote from the concerns of daily life, in the first case Latin, in the second, abstract mathematics.

This book is an attempt to make sense of theoretical physics outsiders within the context of a society that is enriched, enchanted, and awed by science, but also one that in some respects is intimidated by this force. What does it mean that a man in a trailer park feels the need to reinvent physics from the ground up? What does it mean for the man himself, and for society at large?

These questions form the plan of the book. In Part One, I consider what "outsider physics" is. Who are the people doing it? How do they do it? And for how long has this been going on? Our time is not unique for, as we shall see, the mainstream of physics has long been shadowed by a wild and woolly fringe. In Part Two, I look at the case of Jim Carter, the Leonardo of the field. Under what circumstances do Jim's ideas evolve? In Part Three, I attempt to give some insight into the phenomenon as a whole. I hope my conclusions are provocative and, as befits the early days of any field of study, I am sure they can be improved on.

At the start of the twentieth century, collectors and scholars of art began to take seriously what has come to be called the arena of "outsider art." It was clear from the beginning that the terms of engagement with this subject must be different from those usually applied to mainstream art. How should "artists" who operated outside the academies and salons, without formal training, be treated intellectually? How could their achievements be assessed, and what systems of value might be placed on their work? My project is a parallel one: How can we interpret, how can we assess, and how can we understand the "aberrant" body of work that outsider physicists produce?

Part One

OUTSIDER SCIENCE

We have to articulate ourselves. Otherwise we would be cows in the field.

—WERNER HERZOG, *BURDEN OF DREAMS*

Chapter One

UNDER THE HOOD OF THE UNIVERSE

AN HOUR SOUTH of Seattle off the I-5 interstate lies the townlet of Enumclaw, a potpourri of aging weatherboard cottages and unruly gardens nestled within the vast forested landscape of Washington's King County. Huge bushes of rhododendrons bloom in the spring, and much of the year rain seeps incessantly from the sky. You don't see rainbows much around here, but the land feels rich with potential.

Farther to the east, somewhere in central Washington, lies the infamous Mel's Hole, a mysterious pit that is rumored to be bottomless and from which fifty-year-old radio broadcasts are said to ensue, along with weird ethereal lights. Local legends swirl around Mel's Hole: Some suggest it is a warp in the fabric of the spacetime continuum, perhaps concealing the entrance to a wormhole; others propose it is an alien portal through which extraterrestrials are visiting the earth. Jim Carter, a native King County boy, once considered mounting an expedition to search for Mel's Hole, the precise location of which remains unknown. Like many of the region's residents, Jim had heard stories of unexplained phenomena in the area and had wondered what might be behind it all. His own theory of physics denies the existence of a spacetime continuum and does not allow for faster-than-light travel, so in Jim's vision of the universe there are no possibilities for aliens to be mingling among us. As someone who has dug some impressive holes himself, his interest in Mel's Hole was purely geologic: What fascinating things might we learn about the core of our planet from a descent into its depths?

In King County, the holes are of a more mundane variety. This used to be coal-mining territory and in the nearby towns of Black Diamond and Franklin mining was a mainstay of a thriving local economy. The mining companies are gone now but if you go for hikes in the surrounding forest you often come across disused mine shafts grown over with blackberries. Every now and then somebody falls into one, and signs urge caution to the unwary. With the collapse of the mining industry, the region's fortunes took a turn for the worse, but in the 1990s the outflow of a new industrial revolution began to generate renewed hopes for prosperity. A decade earlier, Bill Gates had set up Microsoft headquarters an hour's drive to the north, and as property values near Seattle soared, the new urban elite began to creep southward down the highway, gobbling chunks out of the forest with their expensive gated communities.

The sprawl has not yet hit Enumclaw, and away from the highway today there are few signs of the software giant churning away up yonder. Turning off the I-5 toward Black Diamond and Enumclaw, one immediately moves into a quieter and more feral space—paint peels off clapboards, porches sag, the forest encroaches on gardens, and as if by some immutable law, cars accumulate in front yards. North of Enumclaw's official perimeter, the Green River Gorge Road winds a path of lazy two-lane blacktop through mile after mile of regrowth forest. On either side of the road, dense walls of Douglas firs stand as a testimony to the perpetually soggy weather, as if what falls from the sky is compelled to shoot upward again. Hundreds of thousands of acres of fir trees pile across the hills, a verdant repetitiveness that were it not natural would almost seem computer-generated. Here and there swaths of forest have been logged by timber companies, leaving brown wounds in the great green shim.

Driving through this landscape of monotony induces a meditative effect. With each passing mile, the promise of silicon seems to recede a little further and one passes back into the arms of something more primal and protected. You might be inclined to call the journey boring if you needed to get somewhere in a hurry, but for Jim Carter, who loves nothing more than a wet Washington afternoon, speed is rarely a critical issue. Which is not to say that Jim doesn't like driving fast—he has been known to drive from Enumclaw to Los Angeles in a single eighteen-hour stretch and has spent more time behind a wheel than most people would deem seemly. It's just that Jim is by nature an unhurried man who generally refuses to rush. Time, like space, he regards as his dominion. Out here there seems to be more of both, and in the vastness of this arboreal continuum Jim has carved out a private little rupture in the fabric of reality, a tiny quirky cosmos operating according to its own peculiar laws.

Located on the border of Enumclaw and Black Diamond, Jim's property sits on the rim of the Green River George, a geologic feature described in local tourist industry brochures as "the last great wild place in Western Washington." Known formally as "the Green River Gorge Resort," the Carters' land overlooks a particularly spectacular stretch of the river with cliffs descending two hundred feet to white water surging below. On the floor of the canyon giant boulders sprout in clumps, forcing the water into rapids, while on the lip of the gorge sits the family house, just next to a waterfall. From the Carters' back porch a 180-degree panorama affords a breathtaking view of canyon and forest and sky. The water that has formed the gorge spills off the Cascade Range to the east, pouring through the chasm crystal clear and tinged with trace minerals so that in sunlight it sparkles an ethereal shade of green. It is so clean that the Carters sell rights to a mineral water company whose tankers turn up weekly to siphon it off by the truckload.

Approaching the Carters' property from Seattle, on the northern side of the river, one hears the gorge before one sees it. The sound of rushing water filters through the forest like a gently rising wave, then suddenly the fir trees end and the road opens onto a bridge with the gorge gaping below. Hundreds of millions of years are compressed into this picture-postcard scene. During summer the river shrinks to a stream, and it seems hard to believe that something so modest could have created such a mighty rift, but in winter when rains are in full force, the torrent can rip cedars from the bank and carry away the gigantic logs that fall from above.

Man's response in the face of this majesty has been surprisingly sensitive; erected across the chasm is an elegant steel-and-concrete bridge, its arches complementing the natural cragginess of the land. On one side of this bridge is the world—Black Diamond, Seattle, and Microsoft—on the other side is Jim's world, the idiosyncratic domain of Circlon Synchronicity. On the northern side of the river, universities issue degrees and graduate students beaver away at Ph.D.s, expert committees vet articles for academic journals, and the laws of relativity reign supreme. On the southern side of the river lies Jim's Absolute Motion Institute, with a research population of one, a degree tally of zero, and a publication record unblemished by peer review. Between these two worlds is the chasm, the physical embodiment of a rift that I would gradually come to realize must also be taken on a metaphoric level.

For the casual driver, the bridge across the Green River Gorge offers a chance to slow down and admire nature's work. This far from the highway the roads are pretty clear, and on summer days motorists often park in the middle of the causeway to drink in the view. You can see them from the balcony of Jim's house, which stands high up on the southern lip of the gorge overlooking the bridge. The motorists drop sticks in the river and shout into the chasm to catch the sound of the echoes off its walls, then they drive off to resume their business in the direction of Enumclaw and Mt. Rainier. Some of them are heading for two nearby state parks, the Kanaskat-Palmer and the Hanging Gardens, two popular east Washington destinations for hiking and swimming. The savvy ones know that the best place to swim around here is right below them on the stretch of river that borders the Carters' land.

Across the bridge, one enters a miniature kingdom. On this south side of the gorge the trees immediately close overhead, though here there are also cedars and spruces along with the previously ubiquitous firs. On the right-hand side of the road through a chain-link fence, an observant motorist would notice a pretty section of shaded lawn where a stream has been dammed. Ducks and geese float across the pond. On the left-hand side of the road sits a ramshackle wooden building, a sort of faux alpine lodge painted bright emerald green, with a wood-shingled roof and white-trimmed eaves, its architecture reminiscent of a storybook dwelling. Just past the lodge the road swoops around to the right, and nestled between the trees is a cluster of faded mobile homes and trailers. Around the bend, back on another straight section of road, the Douglas firs take over again and the trailers are soon out of sight and out of mind.

To a casual driver, Jim's dominion would have registered merely as a pretty nice spot on a long and generally pretty road. Aside from the drama of the gorge itself, there is nothing about it to attract scrutiny from passersby. Which is all as it should be to Jim and his trailer-dwelling tenants, many of whom have come here seeking a refuge from that wider and definitely harsher world beyond the bridge. For the residents of Jim's world, the bridge serves as a gateway—the entrance portal, as it were—to a small and private universe where flying under the radar is generally seen as a desirable state of being.

To the more than casual observer, the boundary of Jim's world is well marked. It is chiseled on a wooden sign that hangs out the front of the old emerald lodge. In white letters against a green background, the sign announces that one has arrived at the GREEN RIVER GORGE RESORT. Although that is no longer a strictly accurate description, it is not entirely inappropriate, either. From the 1890s to the 1920s, the lodge was the focus of a thriving local tourist industry; revelers came out from Auburn and Kent, and even down from Seattle, to hike and swim during the day and attend dinner dances at night. Visitors stayed in cabins and enjoyed the wild, unspoiled scenery—the trails down to the river, the waterfall, and the groves of native cedars. The scenery today is much the same as it was a hundred years ago, with sword ferns sprouting beneath the trees and spongy masses of lichen clinging to the branches above. Though the sign out in front of the lodge remains, its party days are long since over; in 1976, when Jim and his wife, Linda, bought the place, it was falling apart. The Carters came to renovate and to give the property a new lease on life.

The project they envisioned was to turn the land into a trailer park, and today the Green River Gorge Resort is home to more than a hundred people. The exact population depends on the number of short-term renters, who tend to be more numerous during the summer; most of the tenants, however, are long-term, some of them having lived here for more than twenty years. Their mobile homes and trailers rest beneath their trees, the steel-and-fiberglass shells contrasting with the lush forms of the forest. On many of the older trailers, the distinction between the manufactured and natural worlds is blurred by the incipient creep of moss—out here *everything* eventually turns green.

The Carters have been generous with their land and have placed the trailer sites sufficiently far apart that each one feels quite private. In all, there are sixty-two "campsites" grouped in clusters throughout the property. Potted plants demarcate boundaries, and one particularly neat encampment has a white picket fence surrounding a carefully mowed lawn. For the most part, however, the trailers are old, and many clearly haven't been moved in a long time. Picnic tables extend living quarters into the open air, although privacy is valued here and the past is a topic to be approached with caution. As in much of rural America, personal space is protected by dogs, who seem to be almost as numerous as the people; frequent barking from the ends of chains and the occasional biting serve as reminders that here, as elsewhere, community spirit has its limitations.

In the glory days of the lodge, visitors stayed in cabins, but those have long since disappeared, and when the Carters came to the land they had to begin afresh. All the trailer hookups, all the water pipes, sewer lines, and power lines, Jim has installed himself. Ground had to be cleared, roads made, vast swatches of blackberries cut down. When Jim needed helpers he hired his tenants, and a kind of bartering system emerged: In exchange for labor, workers received a reduction in their rent. Much of the early work was not skilled, just sheer hard yakka. Jim had a method for sorting out who would make a good laborer that became his entrance exam. Of all the jobs that needed doing, none was more demanding or boring than "rock picking." When a new trailer site was constructed, the ground had to be cleared of rocks so the big power lawn mower could operate, and when someone wanted to work, the first task Jim would set was to pick up rocks. Jim would park his earthmover and leave the worker alone; in an hour or two, he would return to see how many rocks were in the bucket. "If someone was a good rock picker, you could be pretty sure they'd be a good worker," he explains. "If they weren't any good at picking up rocks, they wouldn't be much good for anything else."

Equipment also was bartered, and in lieu of rent the Carters have received cars, piping, electrical cabling, air compressors, power tools, and guns. Jim now has so many guns, they are no longer acceptable tender. There isn't much rock picking these days, but there are always more blackberries to cut down. Everything from maintaining the trails to logging the trees that fall down during storms is done by the community, and in many ways the Green River Gorge Resort operates rather like a medieval village with Jim as benign overlord. Master of the land, builder in chief, dispenser of work, he is the man to whom everyone turns when the drains get blocked.

In addition to dogs, the other major indigenous species of the Green River Gorge Resort is the automobile, and almost every trailer site is encrusted with cars. Like the titans of the computer age whose Saabs and Lexuses ply the highway to Seattle, most "Gorgies"—as tenants are known in Carter family parlance—take pride in their vehicles, and through this glade passes an enormous variety of vintage automobiles: old Mustangs, Pontiacs, and Cadillacs; huge dinged-up Chevys and Ford pickup trucks. One young Gorgie couple boasted a beautiful Plymouth Duster while another tenant—a computer engineer who works on deep-sea fishing boats—used to keep a brand-new Datsun sports car under a tarpaulin.

The car culture that flourishes naturally in blue-collar communities across the nation is given an added boost here by the presence of Trader John, the trailer park's longest-term resident. Green River Gorge Resort rules stipulate that no tenant is supposed to keep more than three cars at any time, but Trader John is, well, a trader, and around his trailer vehicles and vehicle parts multiply in great metallic heaps. Here are to be found things that work, "things that *might* work," "things that are going to be working any day now," and things that Jim's wife, Linda, is pretty certain will never work. In recent years, Linda has attempted to impose a rule that any car parked long-term must be shown to run at least once a month, but neither she nor

Jim has the heart to impose such an order. By Linda's assessment, Jim himself would be breaking the rule most of the time. In any given year, Jim owns up to a dozen cars, mostly huge old Chryslers and Cadillacs from the 1950s and 1960s. They are generally to be seen grazing on the lawn by the side of his house like a herd of contented cows. To hear Jim tell it, every one of them is a functioning proposition, though judging by the flaccidity of some tires, that claim isn't put to the test with any regularity. Whatever their mechanical condition, they are a triumph of American dreaming, with the sharklike fins and yard-long hoods that could have been produced only in an age that had not yet begun to contemplate the end of oil.

Jim Carter is a small man, about five ten, and wiry framed. He is one of those men whose hair began to recede early, giving his head a high, domed, sagelike appearance. At sixty-seven he wears his age lightly, and his eyes, now deeply creased, are prone to twinkle in a face that often lights up with amusement. As a well-brought-up country boy, Jim is too polite to laugh outright at his fellow human beings, but his wide mouth frequently lurches into a smirk, as if he is enjoying some private joke. What hair he does have is now gray, and he wears it at a medium-length cut evenly around the bottom like the fringe hanging off a lampshade. In his youth the hair was sun-bleached, and in photos from the 1960s he exudes a cheeky larrikin charm.

In 1976 when the Carters took over the property, Jim was returning to live in the country that was in his blood. He had grown up in Buckley, a tiny hamlet just down the road from Enumclaw that stands within view of Mt. Rainier. Though he had been born in Seattle, when he was five years old his family moved to Buckley and a forty-acre farm. In addition to farming, his father held a full-time job as a handyman at a local school for the disabled; farming was a supplement and a means of survival. His mother, Phyllis, kept an enormous vegetable garden that went a long way toward feeding the family. Jim and his brother, John, grew up milking cows, mowing hay, repairing tractors, and knowing what it meant to be self-sufficient.

Phyllis now recalls that even as a child Jim always insisted "on taking things apart and putting them back together again." Whenever he was given a new toy, the first thing he would do was to pull it apart to see what was happening inside. Like the young Isaac Newton, who spent his youth building wooden windmills and other mechanical contrivances, Jim came into the world with an inherent love of machines. As Phyllis tells the story, Jim was a headstrong child, determined to do things in his own way. Today, at ninety-seven, Phyllis herself is a force, a minute, birdlike woman in full command of her faculties and with no tolerance for "nonsense." A churchgoing teetotaler, she has spent her long life working hard, living quietly, and staying home. "What would you want to do that for?" is one of her frequent phrases, and it seems more than possible that her son's adventurous nature is at least in part a reaction to his mother's awareness of limits.

It was the farm rather than school that got Jim thinking about science. With "the push and pull reality of a farmboy," he "began to understand how the world must work." Using levers and pulleys and gears, he learned how to control power and enough about electricity and magnetism to build his own motors. Hammering nails, drilling holes, welding metal, and shooting guns, he began to contemplate the forces holding the universe together. When he started attending White River High School in 1959, he learned that this was called "physics," and he began to engage with the subject in a theoretical fashion.

At fourteen Jim acquired his first car, and while still a teenager he began to pride himself on his ability to look under the hood of any vehicle and figure out how to fix it. Today, the only car the Carters own that Jim cannot entirely repair is a mid-nineties Cadillac that Linda inherited from her father: "You lift up the hood, and you barely recognize the engine," Jim says, so loaded is it "with microchips and black-box controls."

Whatever machinery came his way, Jim would turn his hand to getting it running smoothly. "It's just been something I've been forced to do by necessity," he says. "When something breaks down, you fix it. Every machine is a little bit different, and you have to figure out each one for itself, but you can figure out any machine."

In his late teens Jim set out to build a car from scratch, a project that pushed the boundaries of the DIY ethos far beyond the norms of usual rev-head passion and would prove to be a harbinger of things to come, hinting at the directions in which he would later go with his physics. Jim wasn't content to simply *make* a car; his vision was to *design* one from first principles, and to that end he set himself the challenge of constructing a steam-powered automobile. For a teenager the goal proved overreaching, but in his twenties Jim realized a part of this plan when he was granted a patent for a steam engine. He called it the INCOBO, for internal combustion boiler. Unlike other steam engines on the market, in which you have to wait for half an hour after turning them on while the water heats up, the INCOBO would "be on all the time." According to Jim, it would not only be convenient but efficient, so that from the point of view of fuel economy, the INCOBO would be good for our planet. Jim didn't have the funds to build his engine, so instead he built a cardboard model that he animated and filmed; his aim was to show the film to investors he hoped might finance the project. As he saw it, funders would be getting in on the ground floor of an automotive revolution.[1]

Jim's quest for a new kind of steam engine offers us a window into the science he would soon begin to articulate. In Jim's theory of the universe, *everything* is mechanical; like the INCOBO, the world he imagines is made up of simple mechanically interlocking parts. As with his engine, none of the parts are complicated and you don't need much mathematics to understand how it works. In this universe, all matter and energy are explained by the mechanics of subatomic particles each one shaped like a tiny circle of coiled spring. Jim calls this form the "circlon," and in his theory everything that happens in the material world can be explained by the ways in which circlon-shaped particles interact. As in an engine, where gears intermesh, in Jim's universe all things happen through the intermeshing of circlon-shaped parts.

Above all, Jim believes that *atoms* are conglomerations of circlon-shaped particles. Here protons, electrons, and neutrons—the basic building blocks of matter—are different variations of the basic circlon form that link together to form a sort of subatomic mesh. Simple atoms like hydrogen are made up from just a few circlons, while more complicated atoms like uranium are composed from several hundred. In this scheme, circlons fit together in a pattern that mirrors the structure of the periodic table, and for Jim, this table (see insert) is nothing less than a

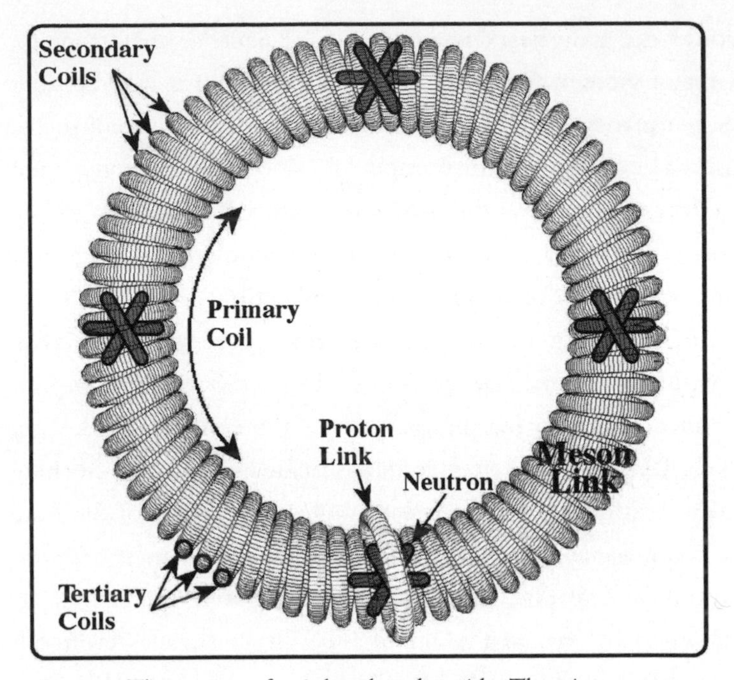

Figure 1. The structure of a circlon-shaped particle. The primary structure is a torus, which itself is composed from a string wound into secondary and tertiary coils. (Jim Carter)

blueprint for what we might see as a series of circlon-based machines.

From the point of view of the physics mainstream, a mechanical universe in any form is pretty hard to accept. To most academic physicists, circlons would seem as quirky as a steam-powered car. I say "would" because no university physicist has read Jim's book. I am the only person with a degree in physics who has. Jim's vision of the universe is literally old-fashioned, for up until the middle of the nineteenth century, most physicists *did* believe our universe was a machine. René Descartes had famously proposed that idea in the early seventeenth century, and for the next two hundred years "mechanism" was the scientific community's reigning philosophy. In the middle decades of the nineteenth century, some of the finest minds in physics were actively trying to articulate mechanical explanations for such basic effects as electric and magnetic forces. James Clerk Maxwell, the Newton of his age, spent decades trying to work out a mechanical explanation for the lines of magnetic force, which he tried to imagine as long thin hollow tubes snaking through space. For much of the history of modern Western science, most scientific thinkers took it for granted that some kind of mechanical explanation would prevail for *all* natural phenomena.

Yet in the latter half of the nineteenth century, a new paradigm worked its way into scientific consciousness and has dominated physics ever since. According to this way of seeing, our universe is composed not from any kind of concrete particles, but from something more ephemeral, what physicists call "fields." The model here is the magnetic field, whose presence can be felt in the region around a magnet by its effect on iron filings. The invisible "field of influence" around a magnet gripped the imaginations of nineteenth-century physicists and finally forced their thinking away from a mechanistic worldview. By the end of the century, mechanism as a philosophy of nature had been relegated to the status of a historical curiosity. As the twentieth century got under way, most professional physicists had come to view the idea of the universe-as-machine on a par with the "phlogiston," which was once thought to explain fire.

Figure 2. James Clerk Maxwell (AIP Emilio Segrè Visual Archives)

In the nineteenth century, physicists had used field theory to explain electricity, magnetism, and light. In the early decades of the twentieth century, the concept was extended to include matter itself, an extraordinary development that took even physicists by surprise. According to the new discipline of "quantum field theory," a "particle" of matter is not a "solid" object at all, but an undulation or ripple in a quantum field that pervades our universe. Here the very concept of "object" is subverted and all our commonsense understandings of that word no longer hold true. "Objects" as we are used to thinking of them don't really exist in the universe of quantum fields. What we are offered instead is a kind of post-object worldview in which the very idea of hard, separate *things* is replaced by a mysterious web of influence. None of this is easy to come at, and as the great quantum pioneer Niels Bohr once remarked, anyone who isn't confused by quantum theory hasn't understood it. Most physicists initially found it hard to accept themselves, yet quantum field theories are supported by equations whose experimental predictions have been borne out to dozens of decimal places. Field theories now underlie mainstream understanding of both matter and energy and are critical to the design of many contemporary technologies, including much of the telecommunications technology we have come to rely on, as well as microchips, a large class of which are made up from "field-effect transistors."

Practically speaking, we are all recipients of the revolution in thinking that the field idea has wrought, and anyone who uses a cell phone or computer has reason to be grateful that physicists have come to understand this enigmatic aspect of our world. Psychically, however, we have paid a price, for the outcome of this intellectual upheaval is a description of our world that few people understand. Fields have become, in effect, the black-box controls of our universe, the theoretical equivalents of the microchip processors that now control our cars. Just as the engine of the twenty-first-century car has become a technological marvel that is inaccessible to backyard mechanics, so the twenty-first-century universe has become an inaccessible wonder, a triumph of theory that can be grasped only by an expertly trained professional class. One way to think about what Jim Carter is doing is that he insists on a universe he can comprehend. As with the old Chryslers and Cadillacs that grace his front yard, Jim demands a cosmos he can figure out for himself.

Of all the things that human beings do, theoretical physics is not one that we tend to associate with amateurs. Since World War II, theoretical physics has become allied to an industry that is wrapped around some of the most complicated facilities our species has constructed—the Hubble Space Telescope, the CERN particle accelerator, the LIGO gravity-wave detector, the Ice-Cube Neutrino Observatory at the South Pole—all of which bear billion-dollar price tags and each of which has huge technical crews devoted to its operation. Such vast experimental enterprises have become essential for the progress of theoretical physics, which relies on verification of its predictions to retain its credibility. Without such machines, theory is in danger of becoming a mathematical game. It is a mark of theorists' enormous success that indeed it now takes so much equipment and so many people to find something that is not already explained. The very abstraction of current theory stands as a testimony to how much physicists have understood, for it is only because they have explained so much that we now find ourselves in truly bewildering territory.

It is not hard to understand why anyone would want to participate in such grand-scale ventures. At the same time, a yearning remains in some physicists' hearts for a smaller-scale, more personal kind of science. Dr. Ken Libbrecht, a physicist I know at Caltech who heads one of the gravity-wave teams, retreats in his spare time from the stage of Big Science to a tiny laboratory where he builds machines to study how snowflakes form. It is something he can do on his own, Libbrecht explains. When he's on holiday from what he calls his "day job" with the LIGO team, snowflakes provide a frontier of research that he can explore by himself. In this field he is the unequivocal world leader, and, surprisingly, very little is known about the physics of ice crystallization. Libbrecht once joked to me that with the papers he works on about gravity waves, the teams of researchers are so large the list of authors may take up more pages than the article. Simply keeping track of everyone's names is a significant challenge for a group leader. With snowflakes the credit line is his alone, and what is more, he is following in the footsteps of scientific giants. Michael Faraday and Johannes Kepler, two of the most important physicists in history, both did research on snowflakes.[2]

Ken Libbrecht isn't the only Caltech physicist who has got a kick from what we might call handmade science. The DIY impulse was also manifest in perhaps the most famous physicist in Caltech's history, the quantum theorist Richard Feynman. Feynman was the scientist who electrified the world on television with his demonstration of why the space shuttle *Challenger* blew up at its launch, killing all the astronauts on board. Those old enough to remember will recall how he dropped a rubber O-ring into a beaker of dry ice and water, causing the O-ring to shatter and thereby explaining how the spacecraft had failed. In 1965, Feynman was awarded the Nobel Prize in physics for his work on quantum field theory, yet in 1964 he set out to perform a task that from the perspective of the scientific mainstream was the equivalent of building a steam engine.

The task Feynman set himself was to derive one of Newton's most important results without using any of the powerful mathematics now available to us. Specifically, he decided to reconstruct one of Newton's key proofs about gravity without using calculus and using only the laws of Euclidean geometry, a branch of mathematics that had been known to the ancient Greeks. Feynman wasn't doing this to advance the state of physics. He

Figure 3. Richard Feynman (Tamiko Thiel, 1983)

was doing it to experience the pleasure of building a law of the universe from scratch. Like Jim Carter with his steam-car project, Feynman wanted to make something important using only the most rudimentary tools. He presented the fruits of his labor to a class of undergraduates at Caltech as one of his legendary Feynman Lectures, and it was an achievement from which he evidently gained an enormous amount of pride. Almost three hundred years after Newton had presented his original proof, Feynman set out to reprise the master's geometric reasoning for his class. Many of the students might already have done the proof themselves with calculus—that is now an undergraduate exercise—and Feynman himself noted that "it's much easier to do with calculus." Some of the students must have wondered why their professor was bothering them with this antediluvian version of the problem. Then Feynman described what he had in mind. "For your entertainment and interest," he said, "I want you to ride in a buggy for its elegance instead of a fancy automobile."

What Feynman set out to do was to prove that if Newton's law of gravity is correct, then the planets necessarily travel around the sun in elliptically shaped orbits. Newton had shown this was true, and his proof played a pivotal role in helping to convince people of the seventeenth century that his gravitational law should be taken seriously. One must bear in mind that at the time, the idea of a mathematical law describing gravity was almost inconceivable. In the seventeenth century, the very notion of an invisible force acting throughout space was bordering on heretical, for it seemed to smack of magic and all the ethereal mumbo-jumbo that the new science was trying to overthrow. Newton understood that his cosmology depended on gravity and that the fate of the new physics rested on his ability to convince his colleagues that what he was saying was real. To make them believe in his law, he felt he had to demonstrate its truth using only the kind of mathematics they would inherently trust. That meant he had to forgo the newfangled *calculus* he'd been inventing and use only the tried and true tools of geometry that even the most conservative mathematicians would accept. Newton presented his gravity law along with his geometric proof about the planetary orbits, in the book that launched his science upon the world. Three hundred years later, Feynman wanted to understand exactly what he had done.

In the preparatory notes Feynman made for his lecture, he

wrote: "Simple things have simple demonstrations." Then, tellingly, he crossed out the second "simple" and replaced it with "elementary." For it turns out there is nothing simple about Newton's proof. Although it uses only rudimentary mathematical tools, it is a masterpiece of intricacy. So arcane is Newton's proof that Feynman could not understand it. That is because in the age of calculus, physicists no longer learn much Euclidean geometry, which, like stonemasonry, has become something of a dying art. Feynman was rather surprised he couldn't follow a piece of scientific reasoning three centuries old, and he seems to have taken that as a personal challenge. Because he couldn't understand Newton's proof, he decided to do a version for himself. The task nearly defeated him, and the result of his work, when it was finally published, occupies close to a hundred typewritten pages. It appears in a marvelous book called *Feynman's Lost Lecture*, by Caltech physicist David Goodstein and his wife, Judith Goodstein, a former Caltech archivist.[3]

Most of Feynman's students probably didn't follow his proof either, but he knew he could expect their applause, and he ended his lecture with a flourish like the showman that he was. When he had found that he couldn't understand Newton's proof, Feynman could have abandoned his idea for the lecture for no one knew he was planning to give it. But he could not let the challenge go, and it is evident from his notes that he derived a tremendous amount of pleasure from his "ride in a buggy." For a man who would soon be granted the highest honor in science, it was a DIY triumph whose only value was the pride and joy that derive from being able to say, "I did it!"[4]

Feynman apparently devoted a fair bit of time to the "buggy" exercise, yet when all is said and done, he was a man on his way to a Nobel Prize, and at the end of his lecture he made a remark that very likely resonated with his Caltech audience. "One should not ride in a buggy all the time," he told the class. "One has the fun of it and then gets out."

The question that stands at the heart of *Physics on the Fringe* is: What happens if one doesn't have the option of getting out of the buggy? To extend Feynman's metaphor, what happens if a buggy is the only form of transportation available? For Richard Feynman, a highly trained theoretician with an exceptional gift for mathematical abstraction, a joyride in "a buggy" was a pleasurable diversion; while he manifestly enjoyed the experience, he did not have to rely on a handmade vehicle for his serious travel needs. For that purpose, he had access to what we might call the "Ferrari" of quantum field theory. We may go even further, for Feynman and his fellow theoretical insiders have access to an entire fleet of fancy automobiles outfitted with all the black-box controls that twentieth-century physics has been able to deliver. These indeed are the men and women who design the black boxes of contemporary physical science.

As a Nobel Prize–winning theoretician, Feynman had the keys to the scientific equivalent of the executive garage. But what if one doesn't have the keys? What if one cannot access the "fancy automobiles"? Metaphorically speaking, with respect to theoretical physics, that is the position of the majority of people on earth. From the perspective of a contemporary theoretical physics insider, what the rest of us understand about the workings of the universe is the equivalent of riding on a bike. Popular physics books keep telling us this, insisting that most of what we believe about our world is outdated, outmoded, or wrong. While the rest of us putter along with our quaint old ways, academic theoretical physicists have been constructing a fleet of evermore complicated Ferraris and Lamborghinis. These are beautiful, impressive machines but few of us can ever hope to "ride" in them ourselves. What unites Jim Carter and other physics outsiders is the belief that the vehicles they build in the privacy of their backyard "laboratories" are just as legitimate as the insiders' elegant automobiles.

Chapter Two

COUNTERPART UNIVERSES "EXCISTING"

"OUTSIDER" THEORIZING IS not something that usually stands at the forefront of a science writer's consciousness. Along with most scientists, science journalists tend to dismiss such people as "cranks," and within the world of science writing this is not exactly a career-enhancing topic. I first heard about the phenomena of physics outsiders soon after I completed my undergraduate degree at Sydney University. It was the early 1980s, and I was beginning to consider whether to go to grad school. A physicist I knew happened to mention in passing that he had received something oddball that week, something that, as I recall, had gone straight into the bin. My reaction then was dismissive, like that of my professor friend. University science training tends to be demanding, and the idea that a person with *no* training could have found the answer to an important scientific question tends to sit rather badly with someone who has just slogged through final-year exams. At the same time, I was intrigued. During my years as a physics student, it had never occurred to me that one could dream of following in Einstein's footsteps unless one had—as Einstein did—actually gotten a degree. What sort of person imagined he or she could succeed as a theoretical physicist *without* the formal training? What sorts of ideas might he or she produce? Behind my umbrage a part of me wanted to know more, but it would be more than a decade before I encountered such a theory myself.

In the meantime I decided not to go to grad school, and I began to forge a career as a science writer. At the time in Australia there was no such thing as a science journalism degree—there were barely any venues for writing about science—and I made my way in my chosen profession by my own rather idiosyncratic path. My first regular journalistic job was writing a monthly science column for a fashion magazine. I engineered the job for myself with the belief that communicating science to women was an important social goal, and for ten years I kept it up, working consecutively for three different magazines, including *Vogue Australia*. It was the first magazine, however, that allowed me to do the most innovative work, and despite its rather insipid name—*Follow Me*—under its auspices I was able to take on some of the most exciting assignments of my career.[1] I was privileged indeed to be one of the few journalists to privately interview Stephen Hawking. In 1986, I spent an afternoon with him in his office at Cambridge University that resulted in a six-page article for the magazine literally posted between photo spreads about cosmetics and coats. Much of the afternoon we spent talking about *time* and Hawking's radical ideas about this subject, all of which became the basis of my article. Eighteen months after the story came out, Hawking published his book *A Brief History of Time* and became an international celebrity, thereby making interviews a good deal more difficult to obtain.

It was through my work as a science journalist that I began to encounter outsider theorists directly. Aside from Jim, my first contact came in 1994 in response to an article I had written for an American science magazine. By this stage I was living in Los Angeles, and one day as I was trying to focus on a new piece of writing, the phone rang. I was behind schedule and under the gun, as journalists typically are, and I almost didn't answer, but the forces of procrastination moved my hand and I found myself on the receiving end of a long outpouring about the True Nature of Reality. My caller claimed to have found "the key to the universe." Millions of dollars were at stake. It was something to do with energy. There was a patent pending and a diagram involved. If he would fax me a peek, I promised, I wouldn't reveal a word until the patent application had cleared. Sadly the fax machine remained silent, yet a new chapter in my life had inadvertently opened. Writing in an Australian newspaper a few months later, I mentioned my mysterious caller, and several weeks later I received in the post a theory from a man named Peter Jobson in the Sydney suburb of Sylvania Hills. Mr. Jobson also had discovered the key to the universe in something he called "the harmonic gate." He described his ideas in a handwritten letter exquisitely drawn up with a fountain pen and illustrated with delicate spidery diagrams. Along with Jim's theory, this was now the third outsider's work I had seen, and I began to wonder how many more of them there might be. How would I get to them? How would *they* get to *me*? The only way forward was to wait and see.

That any kind of collecting must necessarily be ad hoc was brought home to me by the unlikely set of circumstances through which I had come across Jim's work. When he had sent out his original book announcement, I was not on his list of recipients; that had included only famous physicists and scientific booksellers.

At the time I was working on a book of my own, and had been buying a lot of texts from one of L.A.'s leading used-book dealers. One day I was in the store picking up a set of the works of Descartes. I was happy with my purchase and about to leave when the proprietor offered me a package. "This came the other day," she said. "I was about to throw it into the bin, when I thought of you."[2]

After opening the envelope, I saw Jim's order form, the little yellow one with the sense of humor, along with a wall chart he had made showing his circlon-based version of the periodic table. The chart was gorgeous, a symphony of psychedelic hues, which seemed to have been done by a calligrapher on acid. I thanked the bookseller for what seemed a fairly eccentric gift and went home to peruse my treasures—Jim Carter and René Descartes. I never did get around to asking what prompted the dealer to think of me above her other customers, because it did not then occur to me what a significant event this chance encounter would turn out to be in my life. It rather shocks me now to realize that this precious package could so easily have been consigned to yet another bin.

Fifteen years after this encounter, there are two long shelves in my office along with several large boxes devoted to the subject of outsider physics, more than one hundred theories in all. I have not gone out of my way to acquire them, except in the case of Jim's work, which I have assiduously collected. Somehow through the channels of my life they have found their way to my door.

As a collection, the theories on my shelves form a tremendously disparate group, and there is almost no general rule to describe them. Some of them are entire books, some are single articles. Some have been professionally printed, others are typewritten. A small number that I especially value are handwritten, although Peter Jobson's is the only one done in India ink. Some go on for hundreds of pages, although most of the authors seem able to get their points across in a few dozen sheets. Some of them consist almost entirely of words and read more like pieces of literature than articles intended for publication in science journals, but most are scattered with simple equations, and some of them—like Jim's work—are dense with formulae and tables of numbers. Many contain illustrations, which, these days, are usually computer-generated.

What is most surprising is the eclectic range of genres that they come in. While most are presented soberly with the hope that they might be published in an academic journal, there are some less obvious options. One work I especially admire is a book of poems in which the author presents his ideas about space and time. Another is a dreamy personal ramble that flows in a stream of consciousness from the paddocks of Murwillumbah in rural New South Wales to the killing fields of Ohama Beach, offering us along the way a new theory of gravity. For sheer originality of style, the oddest theory I possess is presented as a fairy tale, complete with a little red-haired maiden. In the story, a boy named "Moment" takes a ride on a "Wave Train" in order to encounter the warping of time that Einstein described in his special theory of relativity.

Theories endorsed by the scientific mainstream will usually be sent to journalists like me as formal journal preprints. Theories from the outside tend to arrive fresh off the typewriter. At least that's the way they used to come before word-processing software and page layout programs such as InDesign gave everyone the power to simulate a professionally published text. These days, the packages that arrive in my mailbox are much more highly produced, and I rather regret this trend, for I have noticed in recent years a creeping homogeneity. The works I value most are the ones that are handmade, with all the quirky power of the authors' uniqueness and idiosyncrasy. Even Jim, who retains his visual zest in the digital era, was at his aesthetic peak in the early 1970s when he was cutting and pasting up collages and photocopying the results.

Among the theories in my collection, Jim's work is unique in the totality of his thinking. His work spans the whole of physics and includes a theory of matter, a theory of energy, a theory of gravity, plus a complete account of the creation of the universe. Visually, too, Jim is in a class of his own. I have never encountered another outsider with anything like his aesthetic flair. His books and collages and models, and the many hundreds of illustrations both hand-drawn and computer-generated, all attest to a graphic drive that were it not in the service of science might well have expressed itself in the form of "outsider art." Jim's visual style is of such a caliber that in 2003 I was invited to curate an exhibition of his work at the Santa Monica Museum of Art.[3]

There is very little that unites the disparate range of theories in my collection except for the sense that mainstream physics is badly off course and the authors' beliefs that a few simple ideas will clear up the whole mess. In whatever form they take, outsider theories usually arrive with an explanatory letter alluding to the need for help in getting out the word. Might I, as a journalist, bear the torch of their revolutionary proposals to the *New*

York Times? Might I, as a published author, introduce the correspondent to an editor? An agent? Or to my own publishing company? Singularly, such appeals may be plaintive or defiant. Sometimes they sound a note of harangue. But always through the veil of rejection, a ray of hope will shine: The Truth will win out. *Collectively*, these correspondents represent a retort to the scientific establishment whose sheer number suggests at least pause for thought.

"What is a photon?" asks Irwin Wunderman of Mountain View, California, offering me his answer in *A Unified Wave Theory*. This is one I have selected from my shelf at random; it is a three-hundred-page book typed on A4 paper and photocopied with a Kinko's binding. The author has a Ph.D. in electrical engineering from Stanford University, or so his attached biography says, and I see no reason to doubt him. He has spent forty years honing his ideas. As described in his abstract, Wunderman's theory "attempts to coalesce classical and quantum physics and to derive a numerical origin for particle waves." It is long and technical and extremely dry, and it seems to be an attempt to explain quantum mechanics without using field theory. In his accompanying letter, Wunderman tells me that his ideas are of the utmost significance to humanity and "may influence future methods of information transfer, artificial intelligence, signal processing, equation solving, computer algorithms, space exploration, optics, electronics, and many other fields of science and technology." He is asking for my help in finding a publisher, and he ends his letter with a confident assessment of its prospects: "The scientific conclusions of the work more than justify its publication, and should make it a best seller."

"Questions on the Theory of an Expanding Universe" is the name of a paper sent to me by Denis Nevin of Queensland, Australia, which happens to be where I grew up. The paper is short, the language is bold, the claims are enormous. Mr. Nevin offers an alternative view of our universe's history, and enclosed with his article he also sent me two copies of his book *Intelligent Evolution (A Thermodynamic Unity)*, which has been printed at the Nerang Copy Center. As I read these words, my mind races back to my childhood when I used to attend a school camp at the mouth of the Nerang River. Does Nevin ever visit the Tallebudgera Bird Sanctuary? I wonder. Is the bird sanctuary still there? At Tallebudgera, we children used to pat kangaroos and cuddle koalas. In my memory Nerang is a semijungle. What sort of theory of the universe might issue from this unprepossessing suburb of the Australian Gold Coast?

Nevin's letter gives no sign of an academic affiliation. Unlike Wunderman, he does not offer me a curriculum vitae or mention any credentials. He begins his paper with the premise that the "Big Bang theory accepted by a majority of scientists constitutes the greatest blunder and misinterpretation in the history of cosmology." His own system of the universe is based on something he calls "the Principal Thermodynamic Law of Energetic Oscillation," which implies that "all material in the universe from atom to Man is united in one common and undeviating behavioral pattern." The book he has sent me is the first in a projected series of five volumes that collectively will "explain most major and minor 'human' behavioral patterns with analogous cycles in the sub-species and the outer universe." From there it is on to "creativity—art, music, literature, social organizations—law, democracy, fascism, communism, technology and invention and religion." For the moment, his focus is on the laws of physics. According to Nevin, our universe is compelled toward a "condition of least energy, that which is easiest, fastest, straighter, more logical, better organized, most appropriate, less divisive, lawful, et al., ad infinitum." In his accompanying letter, he acknowledges that these ideas are unorthodox and that effort will be required on the part of the reader to understand what he is saying. To assist me, he offers the following "helpful" summary:

> There is always a FORCE acting—over TIME—to move material of any and every kind into a state of high energy and positional ENTROPY—(randomness, disorder, chaos), but there is always a countervailing force acting—over TIME—to reduce that condition to a state of lower energy and less occupation of space.

The universe might be "straighter, more logical, better organized" Nevin's way; the same cannot be said for his grammar, which seems to be operating in a universe of its own. Cutting through this cloud of ideas, I find he is proposing that the tendency of our universe is not to *expand*, as mainstream physics tells us, but to contract. His theory flatly contradicts most of what we have learned about cosmology during the last hundred years. In the final page of his letter, Nevin summarizes his ideas with six basic laws that seem to be intended as his alternative to Newton's laws of motion. Abruptly after that his confident tone ceases and he seems to falter. "I do feel a bit embarrassed about the above," his letter concludes. "After all, what would a seventy year old former backyard car dealer know about such things!"

The hallmarks of theories like this will be immediately recognized by anyone with a grasp of scientific norms: The very fact that there is no indication of the sender's institution is an immediate giveaway to anyone within the mainstream. Even if a theorist *does* have an institution, it is not usually any organization referenced in the *Chronicle of Higher Education*. In the age of the laser printer, anyone can have a letterhead and a logo—and increasingly everyone does—but it takes more than a trip to Kinko's to register on the citation index of the American Physical Society.

To the trained professional physicist, the code of scientific presentation is complex and nuanced, requiring years of education. Learning to speak and write physics is no less intricate a task than learning to speak and write ancient Chinese. As with all difficult languages, there are fine points that can generally be grasped only after long years of immersion and sustained study. The outsider tends to speak pidgin, a mishmash of languages that is often comprehensible only to himself. If there is one thing that outsiders share, it is a tendency to be unique. That, and a startling inattention to copy editing. Throughout this chapter I reproduce the authors' typography exactly as it appears in the texts they have sent me.

One of the more insistent works in my collection is a book titled simply *Theory of Everything*, by Eugene Sittampalam. It can be bought on his Web site or on Amazon.com. An engineer by profession, Sittampalam tells us on his Web site that he has spent his career consulting for Texas-based construction companies and on projects in oil fields. At night in the field, he would contemplate the stars, and think about man's relation to the cosmic whole. On his Web site he explains the seeds of his thinking:

> How often at night when the heavens are bright with the light from the glittering stars, have I stood there amazed and asked, as I gazed, if their glory exceeds that of ours . . . Surely, I would continue to muse, the framework out there cannot be basically any different to what we have down here. If there *is* a basic difference, then there should be an *interface*—a region where, by implication, the logic on neither side would fully hold. Since the laws of physics, as we know them on "our" side, cannot be compromised, the only rational conclusion is: There is *no* interface between the cosmic-scale structures and the human-scale industrial plants, machines, rigs, pylons, and so on I deal with down here, day in and day out.

Sittampalam voices here a theme that has throughout the history of science propelled insiders and outsiders alike, a belief that nature is governed by a unifying set of laws, and that, therefore, there cannot be places where different laws compete. Our faith in unity is strong, yet what we actually observe is a tension between the laws of general relativity and those of quantum mechanics. Where general relativity describes the laws of the cosmological scale, quantum mechanics tells us about behavior at the subatomic level. Both sets of laws make predictions that have been verified to many decimal places, yet each appears to describe a totally different reality. All "Theories of Everything" are attempts to heal the rift between these vastly disparate views.

According to mainstream physics, the "fundamental" level of reality is the subatomic domain, and it is from this substratum that the realm of human experience is supposed to be derived. In the canonical view, if we humans have a hard time seeing how our subjective perceptions fit into the quantum world-picture, then the problem is not with our physics, but with our conception of ourselves. Like many outsiders, Eugene Sittampalam rejects this displacement of the human self from the center of validity, and insists that his *own* experience must be the starting point for his understanding of the world. The basis of his own theory is his idea that mass and energy are "one and the same," a position from which he goes on to derive a new concept of gravity, and from there his own "Theory of Everything."

Sittampalam's is probably the most long-winded work in my collection, and it's easy to get lost in its fog of words. Within the book there are very few equations and almost no diagrams, which makes it difficult to see what he is trying to say. One of the ways in which an insider can often recognize an outsider theory is when there is an extreme reliance on words. For some outsiders, like Sittampalam, there is a rather annoying tendency to think that if only he or she explains often enough—again and again and again—everything will become clear. But physics in its modern mode has been constructed around mathematics precisely because equations make it possible to express patterns of form and rhythms of physical behavior for which there are no natural language terms. The equations of quantum mechanics, for example, predict that in certain kinds of materials there will be a discontinuity in the way electrons behave, an insight that led to the invention of the microchip. There was no way to come at this insight by words alone. The absence of equations in an outsider theory is a pretty sure sign to a professional physicist that there is unlikely to be much of interest here.

Although in some ways Sittampalam's work is among the least interesting in my collection, he offers something that as far as I know is unique in the world of outsider science: On the front cover of his book, he announces that UP TO ONE MILLION DOLLARS! will be given to any physicist who reads his theory and demonstrates why he is wrong. Until 2009, when he dropped out of sight, he periodically sent out e-mails detailing his bouts with the establishment and the way they had ignored him, new data that supported his ideas, findings that discredited insider theories. He was a powerhouse of dissent. Sittampalam has had no takers for his million dollars, but in 1995 he asked the renowned Harvard particle theorist Howard Georgi to review his theory for a fee. To my knowledge, this is one of the few instances where any outsider has received serious critical feedback from an insider. In Sittampalam's assessment, Dr. Georgi's criticism of his work "proved very constructive, but refutation fell short." Georgi has a rather different view. By e-mail he told me he had spent a great deal of time trying to explain to Sittampalam exactly why his ideas didn't stack up, but nothing Georgi could say would budge him, and in the end Georgi conceded defeat. Yet that was by no means the end of the story. Georgi soon received a second check, this one for $25,000. Sittampalam intended it as "an incentive for anyone at the entire physics department at Harvard to show at least one instance where my model would fail." The person had only to satisfy Georgi that Sittampalam's theory was wrong in order to collect the reward.

Months passed in silence. None of the Harvard faculty took up the challenge. Sittampalam asked that the check be returned.

In response Georgi wrote back: "I have no intention of cashing the second check, but I had intended to keep it as a memento of this rather unusual experience. I hope you will not object." Sittampalam did not object. On the contrary, he "considered it a subtle compliment, coming from a top theorist and authority in the field of modern physics." As he explains on his Web site, "Perhaps the great visionary had a gut feeling that one day that slip of paper would be an invaluable collector's item!"

Recently I encountered another, more subtle story of frustration. The theorist in this case was the legendary Hollywood film editor and sound designer Walter Murch. Murch worked on sound for the George Lucas films *THX 1138* and *American Graffiti* as well as two *Godfather* films. He is most famous for designing the soundscape for *Apocalypse Now*, for which he won his first Academy Award. He remains the only person to have won Oscars for both film editing and sound mixing. I have long been somewhat in awe of Murch, for he edited and designed sound on *The Conversation*, Francis Ford Coppola's masterpiece reflection on sound itself and its ability to both connect us to and alienate us from one another. The hero/antihero of the film, Harry Caul, played by Gene Hackman, is an eavesdropper who designs and builds instruments to record other people's conversations. One of his tapes becomes the centerpiece of a complex murder plot, all of which is presented through the filter of Harry's truncated life. Murch also has edited mainstream blockbusters, including *The English Patient* and *Cold Mountain*, and has been a pioneer in the development of sound-mixing equipment. Like Harry Caul, he is a technical audio genius. Nonetheless he has found the time to work on his own theory of physics.

I first heard about Murch's theory through friends in the

film industry, and in 2010 I had the chance to hear him talk as part of a series of lectures at Occidental College in Los Angeles. Were it not for the topic, there would have been nothing unusual about a man with two Oscars powerfully commanding a stage in front of several hundred well-educated art aficionados. Murch, who is well over six feet tall, makes for an imposing presence: Slim, elegant, dressed in black, he gives the kind of performance one might well expect from a Hollywood prince. In keeping with his profession as a sound designer, he has a wonderful voice—deeply resonant and full of quiet authority. He had come to the event armed not only with a theory, but with an incredible PowerPoint presentation full of diagrams and graphs and tables of data. I can say without hesitation that this was by far the most professionally presented talk I have seen from any outsider, exceeding most in coherence by many orders of magnitude.

Murch's theory is not actually a new theory, but rather an extension of an idea that dates back to the eighteenth century known as the Titius-Bode law, or often just Bode's law. Anyone who has studied astronomy knows about this law, and it has been controversial for two hundred years. The essence of Bode's law states that the distances of the planets from the sun follow a simple numerical formula. When it was first published in the eighteenth century, Bode's law was a reasonable approximation of the distances of the then known planets—Mercury, Venus, Earth, Mars, Jupiter, and Saturn. There was, however, a gap in the predicted sequence between Mars and Jupiter. At the time, all this was seen as an interesting curiosity of no great importance; then in 1781 Uranus was discovered, and its position also fitted neatly into the Bode series. Johann Bode himself urged astronomers to search for a planet in the missing position between Mars and Jupiter, and they subsequently discovered Ceres, a gigantic asteroid, in approximately the right place. But in 1846, Neptune was discovered and its position didn't fit Bode's formula at all. When Pluto was discovered in 1930, it *did* fit into the sequence, but in the place predicted for Neptune. Since then, other planetary-scale objects even farther out haven't fitted the law, at least not according to mainstream astronomers.

Most astronomers throughout this saga have indeed rejected Bode's law. Carl Friedrich Gauss, the foremost expert on the planetary system in the late eighteenth century, argued that it was nothing more than a haphazard coincidence. At the end of the nineteenth century, the American mathematician and logician Charles Sanders Peirce argued that Bode's law was a case of fallacious reasoning. Nonetheless, it has continued to resurface. Every generation or so it comes back into fashion with a new set of champions, most of them outside the academy. Today it is having another resurgence. Indeed, it is such a popular topic now among fringe astronomical theorizers that according to the Wikipedia entry on the subject, "the planetary science journal *Icarus* no longer accepts papers attempting to provide new 'improved' versions of the law."

Bode's law seems to have gripped Walter Murch's soul. Since 1995, he has been amassing what he sees as an ever-expanding set of data in its favor. He has extended the idea to include not just the distances of the planets from our sun, but also the distances of each planet's moons from its parent body. Lots of other astronomers who have looked at this data don't interpret it as verifying Bode's law. According to an article in the astronomy club magazine *FirstLight*, the moons of Jupiter and Uranus actually follow a different law. Murch also looks at the case of extrasolar planets, those orbiting around other stars, and again he sees evidence of the Bode pattern. Such data remains controversial because astronomers are still trying to get precise measurements of these extremely distant bodies and any such figures must at present be seen as provisional. None of this deters Murch, and in his L.A. talk he systematically addressed professionals' reservations and refuted them, piling up his own analysis with ever more detailed PowerPoint slides showing tables of facts and figures. It was a stunning exhibition of DIY research, and it was clear at the end of the hour that he had a lot more slides he could show.

At the reception after the event, I sought Murch out and asked him what he was hoping to achieve. He patiently explained that the evidence he has amassed was now so overwhelming in favor of Bode's law that theoretical physicists could no longer dispute it. It was time for them to start looking for the physical process that was causing it, he said. Did he have any ideas about what that process might be? I inquired.

I hadn't really expected an answer. I'd assumed he was stuck on the law itself and its resonances with musical scales—which is another fascinating part of the story—so I was startled when he replied. Indeed he *did* have an idea. He believed that Bode's law signaled an anomaly in the gravitational field of spacetime and that there was something here that needed to be explained by general relativity. He was genuinely perplexed that physicists weren't taking up the matter. Like Eugene Sittampalam, he felt he had countered all the detractors' arguments, and he could not understand why the theoreticians weren't engaging with him seriously.[4]

Murch wasn't the first Hollywood theorist I had met. One of the field's most well-known figures and one of its elder statesmen is Steven Rado, who is now retired but who spent decades as a graphic artist designing Hollywood movie posters. A native of Hungary, Rado was born in 1920 and has survived both a Nazi labor camp during World War II and the Russian invasion of his country. In each case he fought on the side of the resistance, and on November 4, 1956, when thousands of Soviet tanks invaded Hungary, he led thirty-five freedom fighters and their families across enemy lines to safety in Austria. In 1962, he moved to the United States and ended up in Los Angeles. The tumultuous events of his country and his refusal to kowtow to repressive authorities cost him severely in terms of both his education and his livelihood, so it was not until he had settled into his Hollywood career that he had the liberty to pursue his interest in physics. On his Web site Rado explains that from his "early studies of history, philosophy and political science, he became convinced that the gradual intellectual liberation of people is a direct function of the evolution of science and technology and thus depends on the level of understanding of physics." After his experience with two destructive political regimes, his yearning to understand the universe has been inextricably linked to his desire for a just and sane society.

Rado's theory proposes that our universe is shaped like a giant doughnut. The form arises naturally in his theory from the dynamical motions of an all-pervading "ether." *AETHRO-KINEMATICS: The Reinstatement of Common Sense—An Alternate Solution to the Perplexing Problems of Modern Theoretical Physics and Cosmology* is the name of his book. You can buy it on Amazon, and I highly recommend it as an introduction to the contemporary ether scene. Ether theories are by far the most common class of alternative physics theories, and like Jim's work, they take physics back to the pre-Faraday era. Jim's theory is definitively not an ether theory, which has led him to remark that he is "an outsider even among the outsiders." Ether theories are reminiscent of ideas popular in the seventeenth century, and in these models all physical phenomena are explained by motions in a superfine substance called the "ether" that fills the space between atoms and stars. To anyone versed in the history of science, ether theories have a rather anachronistic ring—most of what I have seen in this vein was said in some form or other three hundred years ago and usually more interestingly. Rado's efforts at envisioning reality are by no means as rich as Jim Carter's, but given his life circumstances we can hardly count that against him. In 1999, I spent an afternoon with him and his wife at their home in the Hollywood Hills, and I could not help but marvel at the strength of will and character it must have taken to lead a group of civilians through armed Russian tanks, walking and crawling for twenty-five miles in the middle of the night. It is no wonder he had the chutzpah to launch a single-handed assault on the foundations of science. In 2000, the Web site Net Prophet awarded him its "Nobul Prize for Exceptional Contributions to Truth."

The Internet, of course, has been a boon for outsider theorists, as it has been for insiders. The World Wide Web was created to facilitate communication among particle physicists—the original Web software was written by Tim Berners-Lee at the CERN accelerator facility—but these days it is also a vast repository of unorthodox thinking. As I sit at my computer, a Google search of the phrase *alternative physics* has just yielded 25.8 million results. On the Open Directory Project Web site, alternative theories are classified into twelve basic groups, including "superluminal physics," "space propulsion," "unified theories," and "unproven energy concepts." On another site under the title "Ratbag Antiphysics Rag," there is a long list of papers dealing with faster-than-light travel and other "anti-relativity ideas." Ether theories especially are thriving online. Here are two that catch my fancy: "NEOETHERICS: Visualizing Gravity" by Jerry Shifman from the Sea Ranch, California; and "AETHER (ETHER), GYRONS and the PHOTON" by Frank Meno, who according to his Web site is a biomedical engineer in the Department of Neurological Surgery at the University of Pittsburgh.

I find it fitting that the Internet should provide such a supportive environment for the new ethereans, for what could be less substantial than the great enveloping cloud of the digital ether. Better still is the support the Net lends to those who harbor perpetual motion schemes, the oldest and most enduring alt.physics dream. The fantasy of an eternal source of free energy is very much alive and well today and remains one of the primary drivers for outsider theorizing. Hundreds of books and entire magazines are devoted to this subject. There are conferences and associations. As a general rule I don't collect theories of this type, which belong to the realm of applied rather than theoretical science. A wondrous repository of information about the subject can be found at the Museum of Unworkable Devices, a witty and entertaining Web site kept by physicist Donald Simanek of Lock Haven University in Pennsylvania. Simanek describes his project as "a celebration of fascinating devices that don't work." His online archive houses "diverse examples of the perverse genius of inventors who refused to let their thinking be intimidated by the laws of nature." Some of the devices on Simanek's site are based on gravity. Others use pneumatic principles. Many are based on hydraulics, including an endless parade of waterwheels dating back to Babylonian times. Myriad curiosities are presented from throughout human history and all over the world. One section of the site is devoted to contemporary ideas involving magnetism, supposed variations in the law of gravity, and so-called zero point energy, a mysterious by-product of quantum mechanics. Best of all is the Web site's Hall of Machinery, where you can watch animations: Because there is no friction in cyberspace, these mechanisms really *will* run forever.[5]

The vast range of outsider physics theories is created by a hugely diverse range of theorists from all walks of life. Over the years, I have generally found that any attempt to generalize about them is prone to fail. Engineers make up the biggest class, and I am sure there is a thesis to be written on why this is so. Frank Meno and Irwin Wunderman aren't the only ones with Ph.D.s, and some "outsiders" actually hold academic jobs, some even in physics departments. Once their ideas become public, these people are often pushed out or made to feel so uncomfortable that they leave. Over the past two decades, there have been several infamous cases of this, including one Russian physicist and a bizarre case from France. In the Russian case, Eugene Podkletnov, a Ph.D. researcher in superconducting materials at the prestigious Tampere Institute of Technology in Finland, claimed to have discovered a device that would reduce the effects of gravity. Because of his credentials, this was big news in the science media, but the academic world reacted harshly to Podkletnov's antigravity claim and he was banned from his lab. Before the dust settled, a NASA team in Huntsville, Alabama, began an attempt to replicate his results. The NASA team never came close to doing so and funding for the project was pulled. Some people have claimed that Boeing also secretly attempted to build a Podkletnov-style antigravity machine.[6]

Outsider physics theorists may come from almost any stratum of life: They may be trained scientists like Podkletnov, college English majors like Walter Murch, or high school dropouts. They may be film editors, backyard car dealers, graphic designers, mathematicians, or artists. I have met examples of all these types. I recently saw a talk by one outsider who was a retired California Supreme Court judge. The only rule I can discern is that most of these theorists are men. One notable exception is Dr. Domina Eberle Spencer, a mathematician (now retired) who has been working for forty years on an "electrodynamic" alternative to special relativity. For better or worse, not many women are dreaming up Theories of Everything or new models of space and time, at least not in the mode of contemporary theoretical physics, for, of course, women have been imagining other models of reality since the dawn of time.

Geographically, also, outsiders are diverse. Most of the items in my collection are from North America, and I can see no particular pattern to their spread, except that I do not have much from the East Coast. My hunch is that this is not a coincidence: Developing a theory of physics is a time-consuming task, and if it's not one's professional life, accompanied by a paycheck, then a good deal of time must be devoted for free. Life in Manhattan or Boston is simply too expensive, I suspect, for the serious amateur theorist to thrive unless he or she is wealthy. I know of one such case that I will discuss later in this book—the man in question began his career as an insider, then from the point of view of the mainstream he went off the rails. Outsider theorists are not confined to the United States. In my collection I have theories from Australia and Holland and the United Kingdom. Australia and the Netherlands are well represented because these are two countries where my profile as a science writer is relatively high. In both places my books on the history of physics have been widely reviewed, and with each publication has come a steady flow of postal offerings.

The spread, and spreading, of outsider theorists can be gauged by the membership of an association to which many of them now belong, the Natural Philosophy Alliance. The NPA today boasts hundreds of members worldwide. At the association's annual conference in 2010, an accompanying book of the conference proceedings contained 120 papers. From the United States, more than twenty states were represented, including California, New York, New Jersey, Oregon, Washington, Pennsylvania, Ohio, Georgia, Texas, Florida, and Hawaii. From outside the United States there were authors from Russia, Poland, Slovakia, Greece, Spain, Denmark, Mexico, Germany, Sweden, Turkey, South Africa, Belarus, Ecuador, Canada, India, Italy, Taiwan, Bulgaria, Austria, Serbia, the United Kingdom, Portugal, Belgium, Australia, and the United Arab Emirates. In a brief introduction to this six-hundred-page volume, editor Greg Volk tells us that about one third of the authors hold doctorate degrees, one third reside outside North America, and one third had never previously contributed to an NPA event. In addition to its conferences, the NPA hosts what it calls the World Science Database, a rapidly expanding archive of alternative physics theories. As of April 2011, the World Science Database contained information on more than nineteen hundred "dissident" scientists, more than thirteen hundred books, almost one thousand Web sites, over one hundred journals, and abstracts of more than five thousand papers.

The NPA's *Proceedings* represent a huge step forward for the organization. Sleekly formatted and well laid out, the book is carefully copyedited and uses a standardized format for diagrams and references. The equations are set using MathType, and at least as far as its presentation goes, the whole publication has a formal, professional air. A glance at the contents pages, however, would leave no mainstream physicist in doubt about its substance. Aside from the unorthodox titles of many papers, a high number of titles take issue with relativity theory. One paper is bluntly called "Einstein Is Wrong." The NPA leadership feels justifiably proud of the conference book; it is the organization's attempt to have its own peer-review publication, and authors were urged before it went to press to read one another's work. The exercise signals a new phase in the NPA's fifteen-year-long history and represents to both the leadership and its members that its aims are serious. As someone who has watched the NPA evolve from the days when its founder was hand-typing newsletters, I have been astounded at how much it *has* evolved. Its Web site is now a vast, invaluable archive of this rapidly expanding field. At the same time, the "professionalization" of the NPA makes me somewhat wistful, for I am certain that the desire for academic legitimacy will swamp the very forces that have driven my most intriguing correspondents. It inclines me all the more to value the oddest things in my collection.

One of the most charming of these was sent to me by Timothy McCormack of Murwillumbah in New South Wales. Murwillumbah is not exactly a focal point of academic ferment. It is in a beautiful part of the Australian countryside close to the Aussie equivalent of Woodstock and has long been famed Down Under as a destination for hippies and pot smokers. I very much doubt McCormack is in either of these categories. While there is no information in his letter by which I might identify his profession, there is a hint in his theory that he may be a veteran of World War II. McCormack's work is entitled "MANY RIVERS TO CROSS: A Short Paper on a Universal Topic," though strictly speaking it is not a scientific paper, but a lyrical series of reflections on the nature of space.

Reading this enigmatic work, I imagine McCormack in a classic Australian country-style home, surrounded by flowering wattles. In my imagination our author is inside typing at his kitchen table, in front of a window that looks out onto the landscape. His paper begins in a pensive mood:

> Not so long ago people knew there was only Air all around them. It was hot, it was cold, it blew fast or slow, it was necessary for fire and life, but that was about it. Today we know that it is composed of many gases, all with different properties that are of great value.
>
> I suggest that today's Space is treated in much the same way as yesterday's Air. What is Space? Can it be defined? Is it nothing? Or is it the physical manifestation of something?

What is space? The question has perplexed philosophers and physicists since at least the time of Aristotle and remains one of the fundamental questions in science. McCormack continues: "Einstein said that mass will curve space time. But what is curved space time, and how does it produce the effect called gravity?

"Any theory of the universe should be able to be seen at work in our day to day environment," he writes.

> If we look at the world we should be able to see Einsteinian visions within our reality.
> This is what I see when I look at my world.
> The clock sits on the windowsill. I see a small fixed mass. I see time at a normal rate and I see reality of a fixed size.
> Now I look out the window deep into the distance.
> I see . . . infinite mass in my vision, countryside, hills, mountains motion slowing down with distance or time slowing. mass shrinks to a point.
> It is an Einsteinian vision that I see . . .

How is it that space and time are altered by the presence of *matter*? McCormack speculates that we might think of space by analogy to water. Just as water pressure increases with depth beneath the ocean, perhaps the effects of gravity increase with "depth" beneath the membrane of spacetime. At this point in his paper, McCormack has inserted several hand-drawn diagrams to assist me in comprehending the flow of his thoughts. The first few are simple textbook diagrams illustrating how a star or planet will "stretch" the membrane of spacetime. These are the kinds of images well-known to viewers of *NOVA* programs or anyone who has read one of Stephen Hawking's books. But I had to pause over the final diagram, which depicts an armored tank pootling beneath the water in a tentlike enclosure. According to

McCormack's explanation, it is "a Duplex Drive tank used on D-Day."

The notion of a swimming tank seemed absurd, yet the reference is tossed off so casually that I decided to look it up, and it turns out that in 1944 the British really *did* fit Sherman tanks with inflatable hoods so they could be launched at sea and "swim ashore." Duplex Drives, or "funnies," as they were known, were supposed to support the Allied troops at the battle of Omaha Beach, but most of the thirty DDs launched on that fateful day sank like stones, drowning their crews and condemning the forces on the beach to terrible losses from German gunfire. In McCormack's theory, a massive star is like one of these tanks, dragging its bulk along beneath the meniscus of space.

Another prize item in my collection comes to me from Peter Jobson in Sydney, who sent me his paper after reading an article I had written in an Australian newspaper. His own paper bears no title, and in a brief nine pages it presents a new "quantum theory of the atom." The text has been written with a fountain pen on thick creamy paper, and judging by the neatness of his penmanship, Mr. Jobson is also of an older generation. His letters are curved, his lines are straight, there are no inkblots or crossings out. In this short work he speaks of black holes, Fibonacci numbers, Schrödinger's theory of subatomic particles, Stephen Hawking's theory of gravitational collapse, and much else besides. The text is illustrated with beautiful spidery diagrams highlighted with yellow and orange felt-tip markers. One drawing is labeled "beginnings of molecules," and in the middle there is a caption stating that "lumps form." Jobson appears to be presenting a "Theory of Everything," though it's pretty hard to make sense of it all. This is one of the most impenetrable works in my collection, and after several readings I have no idea what he is trying to tell me about the fundaments of being. What on earth is the "harmonic gate" by which he sets so much store? The term is never explained. In any case, the theory works its way up to human consciousness, and Jobson concludes his paper with this cryptic declaration:

> Man functions (or is capable of) on 3 levels of consciousness. Time and space are properties of his normal waking consciousness. As there is a (veil) "similar harmonic gate" in man we call the subconscious, the field cannot be witnessed.

I am not quite sure what *can* be witnessed here, though I have many times stared at Jobson's text and tried to penetrate his mind. The confidence of his tone, the density of his prose, those final, bamboozling sentences—what does it all mean? As I read those words, I cannot help but think of the final scene from Stanley Kubrick's *2001: A Space Odyssey.* There is the same aura of portent, the same sense of discovery ahead, and the same nagging sense that behind it all lies madness.

Perhaps the most unusual work in my collection is one that was sent to me by Jim Wallen of Denver, Colorado. It is titled, rather prosaically, "ENERGY TRANSACTIONS: The Fundamental Interactions," which doesn't begin to prepare the reader for what lies ahead. "This paper describes the pattern of all energies in the Universe," Wallen tells me at the outset. The "pattern" in question was first observed by him in a graph used to explain electrical induction, which makes me wonder if he is an electrician or electrical engineer. Whatever his profession, insight dawned when

Wallen realized he had hit upon something fundamental to *all* manifestations of energy and which, he perceived, could hold the key for understanding the relativistic effect known as "time dilation." This is the fact that as things go faster, time appears to slow down. At this point I was expecting a dry technical paper, but instead, after a brief introduction, Wallen shifts modes and the rest of his paper is couched in the form of a fairy tale about a boy catching a train. It is a literary trope inspired by Einstein's famous "thought experiments" about light waves.

The hero of Wallen's story is a boy named "Moment," who at the start of the tale stands on a railway platform waiting to catch the "Wave Train." Moment is a surrogate Einstein about to experience the effects of near light-speed travel. Before he can begin, however, he is distracted from an encounter with these ultimates by "a beautiful little red haired girl" whom he spots on a train moving in the opposite direction. His scientist mind takes a backseat to his romantic heart, and as the Wave Train zooms out of the station, he lies down in his berth for a nap, with the red-haired girl populating his dreams.

Oblivious to the laws of love and obedient only to the laws of physics, the train proceeds along its track, "following the pattern of all energy" in the universe. According to the laws discovered by Wallen, its speed increases in such a way that it never quite reaches the speed of light, which Wallen simply calls 1. Here he pauses to explain Moment's limitation: "He doesn't quite get to a speed of 1. He's slightly retarded. He's not *mentally* retarded, he's just not quite capable of fully concentrating on going forward." His mind is elsewhere; "he's thinking about the little red haired girl." Inside the train, the laws of love threaten to outdo the laws of nature, yet a symmetry protects the universe. As

Wallen explains: "If we were told the other half of this story, we'd discover that the little red haired girl moved at the same rate and was thinking about Moment," so that she is being retarded at the same rate he is.

In Wallen's theory, Moment can *never* reach a speed of 1 because some part of the universe is resistant. As the Wave Train zooms on, the tracks begin to split into multiple branches, a concept Wallen has lifted from quantum mechanics, where one interpretation holds that our universe is constantly splitting into multiple copies of itself. In Moment's universe, each time the train track splits, a small part of the universe's energy is lost. "Some of the energies do not move forward," he continues. "The rails are built from the part that won't move. I look forward to hearing from you," he concludes abruptly.

I did not respond to Mr. Wallen's request for feedback, and I have forgotten how this gem of a theory came my way. I *can* report that although Moment and the little red-haired girl are condemned to remain on "opposing tracks" forever, as "they age and become old they begin kissing each time they pass." Although their trains are going in opposite directions, the train tracks are shaped like a figure eight so that they repeatedly pass each other at a central point. Each time Moment and the girl pass each other, their lips touch and matter is transferred from one train track to the other, forming a *detritus*—that's "Latin for lip dust resulting from kissing," Wallen tells us. This "lip dust" builds up over time on the platform, eventually altering the fate of the universe by changing the pattern of the train lines. I believe I have interpreted things correctly here. If I haven't, I hope that Wallen will write to set me straight.

From another literary direction comes a book of poems by a Dutchman named Elie Agur. Untrained as a scientist, Agur recognized that his chances of having his ideas taken seriously by professional physicists were slim. Instead of attempting to write formal papers, he has presented his theory as a series of poems, with notes explaining the science. I do not know if it is a general feature of Dutch outsiders, but from my experience they win hands down for erudition. Agur quotes Greek myths and gestalt psychologists. There are poems about Cézanne's representations of space and Seurat's pointillism, which Agur believes mirror the structure of physical space. Other poems comment on Newton's laws of motion. There are reveries on Bach's fugues, Proust's madeleines, Marie Antoinette's life, and Sigmund Freud's theory of mind. Charles Darwin, Teilhard de Chardin, William James, Frank Lloyd Wright, and Buckminster Fuller all make an appearance, to say nothing of the physicists whose theories are summarized in Agur's elegant notes. Whatever his contribution to the state of science, Agur is a brilliant advertisement for the Dutch educational system.

Among his *Poems of Time and Space* we find "The Uncared-for Energy (A Natural Experiment with Regard to Space-Energy)," which begins with the following stanzas:

Like a mother
Crying over her forlorn children
Calling upon them
To return
From all corners of the earth
To her, back
To her bosom

Fatherly,
John Archibald Wheeler, entreats
Gravity waves to, please, appear.
Why don't you come, you
Grave-pretty waves
From all corners of the universe?
By what are hindered.
Searchers
Lovingly
Have baked some massive cakes,
Expensive,
To feast on detecting you:
That you deserve
For us to observe
You.

You,
Space-energy,
So diffident you are
That few
Thus far have
Extended a call to face or search you.

The John Archibald Wheeler mentioned here is an eminent relativity theorist and the man who coined the term *black holes.* According to general relativity, black holes should emit gravity waves when they interact with stars and galaxies. Billions of dollars are now being spent on gravitational wave detectors—Agur's "massive cakes"—and theory suggests that such waves should also be coming to us from events like supernova explosions. So far no one has detected a gravity wave, and one purpose of Agur's poem is to suggest that we might switch some of the vast resources here to trying to detect "space-energy" instead. According to his theory, space-energy is responsible for all form and structure in our universe and manifests itself in the shapes of plants and in the geometries of gems and minerals. Agur himself can detect this formative power. Blindfolded and from a distance, he can feel its signature with his hands and distinguish one mineral from another, sensually imbibing the power that unites the animal, vegetable, mineral, and cosmological kingdoms. Like many outsider theorists, Agur is concerned with the Whole of Reality, and his aim is to connect the dots between disparate fields of knowledge.

A final example of an outsider theory was sent to me by another Dutchman, Leo Vuyk in Weesp, and this is one of the most elaborate I have seen. Mr. Vuyk had also read something I'd written, and via my Amsterdam publisher he forwarded me a copy of a paper entitled "Suggestions for the Architecture of Elementary Particles and some Universal Consequences." Vuyk's paper is laid out as if it were a chapter from a physics textbook, and my first thought on receiving it was that it *was* a chapter from a book. His diagrams are detailed and highly technical; they purport to illustrate the effects of gravity around black holes and other relativistic phenomena. Vuyk has an excellent grasp of scientific graphics, but his game is over from an insider perspective as soon as one starts on the text, and you don't need to be a physicist to recognize its delightfully loopy character. Here is one of his conclusions:

> Therefore it was an interesting new step, to give the real form to specific particles, and see the universe as a pure mechanical

impuls [*sic*] and bouncing machine, based on only one variable "virgin particle".

Vuyk goes on in this vein for forty-odd pages. "Virgin particles" aren't the only things needing to be given "real form," and if the universe isn't "a bouncing machine," we can only hope that the next time the Creator takes up tools, He/She/It will consult Mr. Vuyk before proceeding. Recently he has been pushing a new theory that describes our universe as a vast "12 lobed Raspberry in a dodecahedral configuration." Actually his is not a universe, but a multiverse, for here there are "six material universes and six anti-material universes." Collectively they emerge from the "fractal explosion" of "an original Giant Virgin Black Hole." Vuyk has promised a book on these ideas, and I much look forward to seeing it.

In one especially dense section of his paper, Vuyk hints at realities yet unseen. "Is there perhaps a counterpart Universe excisting?" he asks. It seems there is a rather dazzling array of "counterpart Universes excisting," at least insofar as they exist in the minds of their creators. Weesp and Murwillumbah are not the only soils sufficiently loamy to produce outsider physics theories.

In a singular moment of self-reflection, one of my correspondents tells me in his letter that he is a paranoid schizophrenic. That is rare in my experience. Not just the letter, but the fact. Most outsider physicists I have met are perfectly sane individuals, but they can be rather repetitive and they do tend to go on on the same parallel tracks. Like Eugene Sittampalam, they are achingly aware of their rejection by the mainstream and they will go on about that nearly as much as they will about their scientific ideas. Few have many really new ideas, and will keep coming back to a few core themes. In this respect Jim Carter is an unusual case; not only has he made almost no attempt to be acknowledged by the mainstream, he fizzles with original insights. As with the trains in Moment's universe, the track he is on has endlessly diverse configurations.

Chapter Three

A BUDGET OF PARADOXES

IF THE SPECIMENS in my collection are an odd and eclectic mix, I have not been the first to encounter this peculiar blend of qualities. No one has spoken more eloquently on behalf of oddball outsiders, or encountered so many of them, than the English mathematician Augustus De Morgan, who during the nineteenth century amassed what to my knowledge is the largest-ever collection of alternative science theorizing. Culled over a period of more than thirty years, De Morgan's archive is proof that outsider theorizing transcends both geographic and historical boundaries. The fruit of his "research"—if such a term can be applied to the ad hoc accumulation of "stuff"—was a book published posthumously in 1872 under the charmingly Victorian title *A Budget of Paradoxes.* To the twenty-first-century reader, the name now seems obtuse, yet in the nineteenth century a "budget" was nothing more than an assortment of things. "The contents of a wallet, a bundle, a collection, a stock" is how the *Oxford English Dictionary* defines the word. A "paradox" we all understand to be a logical contradiction, yet De Morgan wished to stress a rather different meaning. "I use the word in the old sense," he tells us, meaning to refer to something simply "apart from the general opinion." In its old-fashioned usage, a "paradox" was an exception to prevailing views, and the men who held such opinions De Morgan fondly referred to as his "paradoxers."

A "budget" they certainly are. In the 1954 Dover reprint, they fill four hundred pages of closely packed type. Evidently the Dover editors felt a need to spice-up the bamboozling title, for their front cover announces De Morgan's mission in Barnum & Bailey–esque terms:

> 400 classic examples of scientific logic gone haywire, gleefully collected and mercilessly exposed by one of the wittiest mathematical innovators of the 19th century.

I suppose the publishers thought that "lunatics" would sell, especially when filleted by a scientific giant. De Morgan himself was unconcerned with the question of sales, for by the time the *Budget* was published, he was dead. He had been collecting his examples for "a third of a century," and the compilation of the book was one of the last projects in a long and luminous life at the forefront of British mathematics and logic.

The *Budget* was a compilation of columns that De Morgan had written for the London magazine the *Athenaeum*, the literary arm of the elite Athenaeum Club. Primarily a journal of literary and cultural criticism, the *Athenaeum* also maintained an active interest in science, and De Morgan was one of the magazine's most long-standing writers. Later contributors included T. S. Eliot, Thomas Hardy, and Virginia Woolf. In his columns, De Morgan commented freely on anything that vexed his prodigious mind; usually he wrote about things that annoyed or irked or amazed him in some reprehensible way, though his

Figure 4. Augustus De Morgan (SPL/Photo Researchers, Inc.)

tone is light and throughout it all he maintains a sense of wonder at humanity's capacity for diversion.

In spite of the Dover cover copy, only about half the entries in the *Budget* are about scientific and mathematical theories, but in four hundred pages there are plenty enough of these. Paradoxers who are not commenting on science offer alternative ideas about history, theology, philology, linguistics, and a variety of other subjects. As De Morgan's widow, Sophia, noted in her preface, the *Budget* was "in some degree a receptacle for the author's thoughts on any literary, scientific or social question" he deemed fit to comment on. And De Morgan was a man who felt fit to comment on pretty much everything.

In the *Budget*, De Morgan reserves a particular name for theorists of a scientific and mathematical stripe: He calls them his "discoverers," and they seem to occupy a special place in his large and open heart. Like the theorists in my own collection, De Morgan's discoverers are interested in the great scientific questions of their time: What is the origin of the moon? What is the source of energy that powers the sun? What is the nature of atoms? How was the solar system formed? What is light? What is gravity? Some of these are the same questions my own correspondents are engaging with today, though of course De Morgan's men were working in a pre-relativistic and pre-quantum era, so their answers have a very different slant.

Figure 5. Isaac Newton (AIP Emilio Segrè Visual Archives)

It is interesting reading the *Budget* now to see what a vast difference 150 years have made in our thinking about the world, for on almost every subject, the grounds of scientific discourse have changed beyond imagining. Skimming through the book, one is immediately struck by how uncomplicated and intellectually innocent are the theories on display. Where many of the theories in the *Proceedings of the Natural Philosophy Alliance* exhibit labyrinthine complexity, those in the *Budget* can generally be summed up in a sentence or two. Another noticeable difference is that there is no Einstein to rail against, for he had not yet been born, and Isaac Newton stands instead as the incarnation of scientific "hegemony." It is astounding the number of ways in which Newton is proved to be wrong in the *Budget*. The fellow must have been a dunce.

In the *Budget* we are privy only to De Morgan's descriptions of the authors' works, along with whatever commentary he cares to add. We do not get to see the original papers. Some entries take up dozens of pages; others are dispatched with a few lines. Here is De Morgan's description of one work about the sun, a subject on which the author has apparently made a remarkable discovery:

> A treatise on the sublime science of heliography, satisfactorily demonstrating our great orb of light, the sun, to be absolutely no other than a body of ice! Overturning all the received systems of the universe hitherto extant; proving the celebrated and indefatigable Sir Isaac Newton, in his theory of solar system, to be as far distant from the truth, as many of the heathen authors of Greece and Rome. —*Charles Palmer, Gent. London, 1798*

Charles Palmer appears to have been a gentleman of means. After burning some tobacco with a magnifying glass, he had apparently come to believe that he could have done the same with a lens made of ice. What works on earth must work in the heavens, so putting two and two together, Palmer concluded that our sun must be a giant frozen lens. De Morgan dispatches this idea swiftly and moves on to another disputer of Newton. Here is his description of yet *another* anti-Newtonian theory, this one about gravity:

> An inquiry into the cause of what is called gravitation or attraction, in which the motions of the heavenly bodies and the preservation and operations of all nature are deduced from the universal principles of efflux and reflux.

De Morgan restrains from commenting on "efflux" and "reflux," and I am rather sorry not to hear his thoughts on the matter, though I suppose absence speaks louder than any words might have done. Throughout the *Budget*, the existence of gravity is proved, disproved, denied, explained, and extended. It exists. It doesn't exist. It is an illusion. It is a delusion. It is always a confusion. Whatever the nub of the problem, the discoverer is there with a nifty solution.

As with my own collection, De Morgan amassed some literary oddballs. "The Elements of Geometry written entirely without any punctuation marks by the Reverend J. Dobson 1815," reads one entry. De Morgan then proceeds to critique this work in the manner of the author himself, without the aid of a single comma or period the results of which I will leave to the reader to imagine though I hope this tiny sample will be more than enough.

Also included in De Morgan's lineup are amateur mathematicians who apply themselves to questions such as the derivation of pi. "Squarers of the circle, trisectors of the angle, duplicators of the cube, constructors of perpetual motion, subverters of gravitation, stagnators of the earth, builders of the universe," is how De Morgan wittily describes these men. But while De Morgan undoubtedly could see the humor in his subject (clearly the items in the *Budget* were "gleefully collected," as the Dover cover tells us), he was no more "merciless" with his paradoxers than Lewis Carroll was with the white rabbits and talking chess pieces that populate Wonderland. *Alice's Adventures in Wonderland* was itself published in 1865, just a few years before the *Budget*, and its author, Charles Dodgson, was a mathematician who worked in the same field of logic as De Morgan. Like Carroll, De Morgan had a sharp ear for implausible leaps of logic, and the *Budget* is a Wonderland of its own, populated by an equally unlikely cast of endearingly offbeat creatures.

Throughout the book, De Morgan shows brio and flair in critiquing his paradoxers' theses, yet he always maintains a sympathetic ear for the men themselves and never resorts to derogatory remarks. It is as if, like Lewis Carroll, he views himself as a champion of these befuddled souls, who might so easily be targets of derision from supposedly more rational beings. In this respect, De Morgan understood that his intentions with the *Budget* might be misconstrued, and in his introductory essay he insists that no one should read malignancy within: "Many of the things brought forward here would now be called *crotchets*," he wrote. That "is the nearest word we have now to the old *paradox*. But there is a difference, that by calling a thing a crotchet we mean to speak lightly of it; which was not the necessary sense of paradox." Here De Morgan had correctly anticipated how many of his academic colleagues would react, and he did his utmost to cut off derision from the start. This was one respect in which his wishes did not prevail.

In an amusingly insightful introduction to the Dover edition, the philosopher of science Ernest Nagel summarizes the *Budget* as "a survey of . . . the huge debris of intellectual labor which borders the winding path cut by modern science though the jungle of human ignorance." What a vivid description this is, setting up an image of science as a clearway through a jungle and giving us a visual metaphor at once powerful and progressive. Yet like many supposedly positive views from the Victorian era, Nagel's metaphor brings to mind ideas that do not sit so easily with us now. Humans have indeed cut swaths through the Amazon and Congo, opening the way to riches and expanding "civilization," but the end result, as we now see, also entails a good deal of destruction and waste. For one thing, physical forestry generates "debris"—smaller trees that get tossed aside by loggers and are left by the wayside to rot. So, too, when the intellectual loggers of science have done their work there will always be lots of smaller saplings that have been tossed aside in their wake. The *Budget* is a catalog of the debris left behind in the march of nineteenth-century science; it is one man's collection of the scientific establishment's rejects. To use a slightly different metaphor, we might also draw an analogy here with a midden, the piles of refuse that mark many ancient human settlement sites and that anthropologists now see as rich sources of information about how our ancestors lived. The *Budget* is a midden of sorts: It's what the tribe of academic scientists has thrown out. From the point of view of the mainstream, it is literally trash.

De Morgan's attitude to this "trash" is on the whole neutral. The *Budget* "presents no thesis and points no moral," Nagel tells us. De Morgan had collected these ideas and given them space on his shelves, but he did not intend to give us an analysis of what it all means. What we have on show in the *Budget* is an ad hoc assortment of stuff organized in no particular way and given no particular ordering or categorizing. Widow Sophia notes the volume's "miscellaneous and discursive character." No one, however, explained the project better than De Morgan, who wrote a long introductory essay to the book describing how he came to have the items on display. In preparing the reader for his subject, De Morgan used an analogy from the animal rather than the vegetable kingdom, likening his paradoxers to "flies" flitting en masse around the great "elephant" of established science, and the book opens with the following marvelous passage:

> If I had before me a fly and an elephant, having never seen more than one such magnitude of either kind; and if the fly were to endeavor to persuade me that he was larger than the elephant, I might by possibility be placed in a difficulty. The apparently little creature might use such arguments about the effect of distance, and might appeal to such laws of sight and hearing as I, if unlearned in those things, might be unable to wholly reject. But if there were a thousand flies all buzzing, to appearance, about the great creature; and, to a fly, declaring, each one for himself, that he was bigger than the quadruped; and all giving different and frequently contradictory reasons; and each one despising and opposing the reasons of

> the others—I should feel quite at my ease. I should certainly say, "My Little Friends, the case of each one of you is destroyed by the rest." I intend to show flies in the swarm with a few larger animals.

De Morgan did not mean any of this in a derogatory way. The fly analogy was no slur, but rather the reflection of an actual paradox: If Truth is the discoverer's claim, then how can so many of them possess it simultaneously? De Morgan regarded himself as a naturalist in the field, and his goal was to capture the qualities of his chosen species as a *group*.

De Morgan knew firsthand how many paradoxers there were. "I suspect I know more of the class than any man in Britain," he wrote. It is hard to imagine he was wrong. As a columnist for the *Athenaeum*, he was a natural target for people with an idea to sell, much as high-profile journalists are today, and the postman would regularly deliver new theories in the mail. Others he acquired as books that were sent in for review, and to this task he duly rose, though usually not with the endorsements the authors were no doubt hoping to receive. Some theories he purchased at book fairs or estate sales. Others he obtained from friends who found such items "among what they called their rubbish." As with my own collection, no program or curatorial strategy was applied. Indeed, "no *selection* [was] made at all." Whatever happened to come De Morgan's way would be incorporated into his archive regardless of its subject or style or the credentials of its creator. The *Budget* was formed, as De Morgan put it, "by accident and circumstance alone."

In addition to the manuscripts he collected, De Morgan was visited in person by a steady tide of paradoxers. "I never kept any reckoning," he wrote, "but I know . . . I have talked to more than five in each year, giving more than a hundred and fifty specimens. Of this I am sure, that it is my own fault if they have not been a thousand. Nobody knows how they swarm, except those to whom they naturally resort. They are in all ranks and occupations, of all ages and character. They are very earnest people, and their purpose is bona fide the dissemination of their paradoxes."

Half a dozen times a year, some discoverer would come knocking at De Morgan's door, certain that in person he could convince the great professor to see the Light. The next hour would be given over to hearing the paradoxer out. Throughout these visits, a fatherly solicitousness prevailed, the tenor of which may be glimpsed in a marvelous passage in the *Budget* that remains one of my favorites in the book.

"An elderly gentleman came to show me how the universe was created," De Morgan begins.

We may picture the man—small, eager, and earnest, standing on De Morgan's hearth rug with a much-thumbed sheaf of papers in his hand. He is alive with his inner vision. He has the professor's attention. He has but to open the floodgates and a new era in human understanding will begin. His theory is about molecules and how they produce the bodies in our solar system. Here De Morgan takes up the story:

> There was one molecule, which by vibration became—Heaven knows how!—the Sun. Further vibrations produced Mercury, and so on. I suspect the nebular hypothesis had got into the poor man's head by reading, in some singular mixture with what it found there. Some modifications of vibration gave heat, electricity etc.

I listened until my informant ceased to vibrate—which is always the shortest way—and then said, "Our knowledge of elastic fluids is imperfect."

"Sir!" said he, "I see you perceive the truth of what I have said."

Most scientists, unfortunately, are a good deal less tolerant than De Morgan. Next on this particular discoverer's list of potential converts to his theory was one Dr. Lardner, who "would not go into the system at all." "The first molecule settled the question," and Lardner sent the fellow packing. De Morgan offers his sympathies, ending his account with a gentle reproach to his colleague: "So hard upon poor discoverers are men of science," he concludes.

Hard indeed are men of science on humble discoverers. Then as now, the Dr. Lardners have been the norm. Yet De Morgan himself points out an interesting anomaly in the scene above: Although his reaction has been neutral, the paradoxer takes the absence of a refutation as a signal of assent. So used is the poor discoverer to abuse from the scientific mainstream that the least hint of civility comes to him as an endorsement. What he wants, what he craves, is to have his ideas heard. He wishes to share the joy of his discovery, and *any* attention at all he will take in a positive light.

It is no wonder that would-be theorists sought the ear of De Morgan; he was one of the great mathematicians of the nineteenth century. Among his many areas of achievement, De Morgan played a seminal role in the development of mathematical logic, which provides the theoretical underpinning for digital computing. Some of the fundamental laws of logic are known as "De Morgan's laws," and they are now taught to students of computer science. De Morgan was at the forefront of a transformation in the whole field of mathematics, helping to lead the way into a new era in which the very idea of what math is was being reconceived. His attitude to his discoverers was grounded in his awareness of the great shifts taking place in our concept of knowledge itself. The average eminent scientist of the nineteenth century was no less likely to look kindly on a discoverer than the average eminent scientist today. If De Morgan was sympathetic to these men, at least part of the reason lay in his own history, for he saw himself as something of an outsider to the establishment of his time. Although he is now regarded as a top-drawer authority with a set of laws named after him to cement his name in the pantheon, to his contemporaries De Morgan was a maverick who worked in what was then considered a second-tier institution.

London University, or what later came to be known as University College London, would itself become a bastion of prestige, yet when De Morgan started working there, at the age of twenty-two, it was a brand-new establishment and the first of what the British rather derisively refer to as "red-brick institutions." This snide term delineates more recent academic establishments from the elegant old colleges of Oxford and Cambridge, many of which can trace their roots to the Middle Ages. London University was specifically founded in 1826 by a group of liberal-minded reformers as a reaction to the demand made by Oxbridge colleges that higher-degree students pass a theological test. As part of this "test" master's level students had to swear fealty to the beliefs of the Church of England, including the doctrine of the

Holy Trinity. The founders of London University wanted an institution that was theologically neutral. As a student at Cambridge, De Morgan had objected to taking the test and was thus denied the possibility of an Oxbridge career. So when the new university offered him a position as its professor of mathematics, he jumped at the chance. He would hold the job for the next thirty years.[1]

It is rather extraordinary to ponder the fact that a man as brilliant as De Morgan could be denied entry into the academic world because he would not admit to a belief in the trinity of God. It was irrelevant whether or not he believed in God; De Morgan did not think he ought to have to say what he believed on any religious subject. Throughout his life he championed the idea of academic freedom, and even at UCL, a bastion of liberalism, he resigned his position *twice* because he objected to decisions of the university senate that he felt were detrimental to this principle. It is an interesting comment on the changing nature of university careers that De Morgan could be simultaneously a world-class logician, a columnist for a literary magazine, and an activist on behalf of intellectual freedom. It is hard to imagine any mathematician today being so successfully diverse.

Perhaps because of his experience at Cambridge, De Morgan maintained a lifelong antipathy to authority, and on many levels throughout his career he bucked "the system." Lots of great minds resent working in lower-ranked institutions, but De Morgan seems to have relished the situation and went out of his way to thumb his nose at the academic elite. He was particularly dismissive about the Royal Society, the most important scientific organization in the world, availing himself of every opportunity to belittle its members. He used to say that he had nothing in common with these "physical philosophers," and in his work as a mathematician he was at pains to stress that his research had nothing to do with the material world. In short, in the milieu of nineteenth-century British science, De Morgan himself was an upstart proposing wildly new ideas. From the perspective of many Royal Society members, *he* was a paradoxer.

What De Morgan was discovering was a new way to think about mathematics, and he helped to bring into being new branches of this subject. This new math would turn out to have enormous consequences for physics, which from this time on became ever more mathematical in character. Thus the very quality about this science that NPA members so resent traces its roots to De Morgan's time. De Morgan understood the nature of this intellectual shift, and much of his attitude to his discoverers was based on his awareness of how bamboozling it all must seem to non-experts. Discoverers weren't the only ones perplexed—many professionals also had a hard time adjusting to the new landscape of ideas.

To understand what De Morgan was doing, it helps to know a little about the history of mathematics. Prior to the scientific revolution, mathematics had been seen as the study of quantity and form—*quantity* referred to numbers (the things that enable us to count) and *form* referred to that which is described by geometry (planes and spheres and cubes and so on). In the seventeenth century, calculus had been introduced, bringing the notion of change and time into the mathematical arena. Until the nineteenth century, all three concepts—quantity, form, and change—were largely understood via their instantiation in the material world: Quantity was allied with counting objects like sheep and pieces of gold. The study of form, or geo-metry, had historically arisen out of the need to measure land so that property boundaries could be drawn. Change was a quality of motion allied to concepts like the velocity and acceleration of a cannonball. De Morgan helped to bring about a philosophical shift in which our thinking about mathematics became divorced from the physical world, and the subject was reconceived on a far more abstract basis.

With De Morgan helping to lead the way, mathematics now came to be seen as a discipline that need not have anything to do with the material world. In this new way of seeing, numbers and points and lines were now considered as things in *themselves* that need not relate to any objects humans can see or touch. As De Morgan now put it, the only formal requirement for a mathematical system was that it be internally self-consistent. De Morgan summed up this new attitude with the extraordinary assertion that mathematics is the "the science of symbols." As such, it is not *about* anything, other than itself. In a revolutionary chapter of his book *Trigonometry and Double Algebra*, De Morgan instructed his readers to "bear in mind" that herein "no word or sign of arithmetic or algebra has one atom of meaning."

In this new way of seeing, a "number" is not something that enables us to count sheep. It is no more, and no less, than a participating member in a set of relationships that mathematicians call the "integers." Here the 3-ness of 3 has very little to do with the fact that there are 3 sheep in my paddock or 3 coins in my purse: "3" is just one of the integers.[2] According to De Morgan, it is irrelevant *how*, or even *if*, mathematical concepts are manifest in the world. To the mathematician, *as* a mathematician, it matters not if some mathematical concept seems to be physically impossible, like the so-called imaginary numbers. As long as a concept belongs to a logically viable set of relationships, it is a legitimate focus of mathematical study. It was this alarming shift in perspective that helped to bring into being many of the new branches of mathematics for which the nineteenth century is famous: topology, group theory, matrix algebra, and myriad other puzzlements.

If you as a reader are feeling at sea, don't be alarmed. Mathematicians also struggled with these ideas. Yet we are all familiar with at least some of what this has produced. The Möbius strip, for instance, discovered in 1858, is a surface with only one side. The Klein bottle, discovered in 1882, is a bottlelike structure whose inside and outside are one. De Morgan himself contributed to the development of a four-dimensional number system called the "quaternions." What exactly it meant to have a four-dimensional number was mysterious, but the internal logic of the ordinary numbers seemed to necessitate such a thing. Multidimensionality and topology both were brand-new concepts, and mathematicians were beginning to see that anything that wasn't a logical contradiction would *have* to be allowed.

All this formed the background to Lewis Carroll's stories, particularly *Alice's Adventures in Wonderland* and *Through the Looking-Glass*, published just one year before the *Budget* came out. The author of these enchanted fables, Charles Dodgson, was a mathematician at Oxford. Dodgson, also, had an acute awareness of the intellectual changes taking place in his time, and like De Morgan, he contributed to the development of some bizarre new concepts, including matrices. There is a lovely story told about Queen Victoria, who was so taken with *Alice's Adventures in Wonderland* that she suggested that Carroll dedicate his next book to her. Dodgson duly obliged by sending her a treatise on matrices. There was more than a little overlap between the surreal world of Wonderland and the "real" world in which its author lived. When the White Queen says to Alice in *Through the Looking-Glass* that on a good day she "can believe six impossible things before breakfast," she was doing only what mathematicians of the late nineteenth century were gradually becoming accustomed to. The mathematical mind had been freed to play in the field of ideas, and much of its activity *would* henceforth seem absurd to non-experts. De Morgan's kindness to his discoverers was informed by his understanding of how alienating these developments must necessarily be to those outside the tiny world of mathematics professionals.

Mathematicians weren't the only ones going down a more abstract path. Physicists were also turning to newfangled math in their quest to describe our world. And what is so remarkable here is that so much of the bizarre new math did turn out to be manifest in the world. Imaginary numbers, for instance, are necessary for describing the behavior of electrical circuits and light. Quaternions describe mechanics in three-dimensional space. Matrices turn up all over the place, including in the behavior of subatomic particles. The physical manifestation of such "abstruse" concepts is perhaps the greatest mystery in physics and has been the source of enormous debate ever since. Why does nature have recourse to such abstractions? What does it mean that these counterintuitive formalisms are "present" in our world? As the physicist Eugene Wigner would later famously put it, "the unreasonable effectiveness of mathematics" is one of the prime conundrums of existence.

But the price of such abstraction has been a description of our world that few non-experts can hope to understand. De Morgan's discoverers belong to this class, and in *A Budget of Paradoxes* they come across a lot like *Wonderland*'s befuddled bunny—innocents loose in a landscape beyond their comprehension. Just as Lewis Carroll cast a sympathetic eye toward his rabbit, so De Morgan made it clear that is not the discoverer's fault if the laws of nature seem to be against him. In the new landscape of modern physics, we are all of us in a real-life Mad Hatter's tea party, and in many ways the world we inhabit confounds common sense.

The discoverers who came to De Morgan's door were fired up with enthusiasm, but for the most part they seemed unaware of the monumental developments taking place in the science of their time. The nineteenth century has justifiably been called "the heroic age of physics," and major developments were going on in almost every branch of this science, many of them drawing on the new mathematics. One of De Morgan's discoverers had a theory about molecules in the sun; well, in 1814 Joseph von Fraunhofer discovered dark lines in the spectrum of sunlight and initiated spectroscopy, a science that actually does tell us the molecules present in stars. Several of De Morgan's theorists had ideas about heat. In 1822, the French mathematical physicist Joseph Fourier published his landmark theory about the mathematics of heat flow that continues to be a source of value to this day.[3] In 1865, James Clerk Maxwell introduced a mathematical version of Faraday's ideas about electric and magnetic fields and used his equations to predict the existence of radio waves. Based on Maxwell's equations, Heinrich Hertz set out to look for these waves, which he detected in 1886, thereby inaugurating the age of telecommunications. The new mathematical approach to physics allowed scientists to develop a sophisticated theory of thermodynamics and thereby helped in the development of steam engines. During De Morgan's lifetime, physics informed by the new math began to impact on Western culture in ways that had been inconceivable in any previous age, and it now began to take a place on the center stage of an emerging industrial economy.

If discoverers were interested in the great questions of science, why didn't they go to night school or read books to learn about the new science being done? Since at least the 1830s, physicists *had* been trying to convey their ideas in popular lectures and books. Here too De Morgan played a role. The same body of reformers who founded London University also started an organization to help educate the public on a wide range of subjects, including science. The Society for the Diffusion of Useful Knowledge, as it was called, was formed specifically "to spread scientific and other knowledge by means of cheap and clearly written treatises by the best writers of the time." For the society De Morgan wrote a monograph on calculus. Most important, the S.D.U.K. published *The Penny Cyclopaedia*, one of the English-speaking-world's first encyclopedias and one of the great early works of public education. For this ultimately vast project, De Morgan wrote more than seven hundred articles.

Despite the abstraction of De Morgan's own field, he was deeply committed to helping the public understand the new developments in mathematics and science. So why, then, to put the question into his terms, is it that the magnificent "elephant" of mainstream science was being shadowed by a cloud of "flies"? This is a question that the *Budget* does not attempt to answer. In his book, De Morgan raises the issue of motive only rather obliquely when from time to time the subject of hoaxing comes up. As he notes, there have always been charlatans who have used science in pursuit of shady financial schemes—think of the fudging of data that has recently plagued the pharmaceutical industry. Yet a desire to deceive cannot be an explanation for discoverers because the vast majority of these people believe their ideas to be true. From the perspective of the insider, what the outsider believes, he believes *too much*. It is an *excess* of integrity rather than a lack of it that drives the outsider theorist. The discoverer's fault is not that he attempts to con, but rather that he is taken in himself by his imaginings. It is his very sincerity that is the problem.

All this De Morgan perceived with sensitivity and sympathy, and in defense of his discoverers, he declared rather pointedly that if the fault lay anywhere in this regard, it was with the accusers. "The easy belief in roguery and intentional imposture which prevails in educated society is, to my mind, a greater presumption against the honesty of mankind than all the roguery and imposture itself," he wrote. "Putting aside mere swindling for the sake of gain," De Morgan found "very little reason to suspect willful deceit" from any of the men who came knocking at his door. From my own encounters, I concur—there is no one more earnest than a discoverer. Inasmuch as De Morgan proposed any psychological drives, he pointed the finger at credulity: "My opinion of mankind is founded upon the mournful fact that, so far as I can see, they find within themselves the means of believing in a thousand times as much as there is to believe in judging by experience."

In the *Budget*, De Morgan exposed his paradoxers' methods and conclusions to critical analysis, but for the most part he left the issue of their motives unaddressed. And so we come back to our question: Why does anyone feel the need to reinvent physics for him- or herself? Pretty much every physicist who has commented since, has, like De Morgan, attributed the phenomenon to gullibility. At least that's what the kinder ones say. Those of a harsher stripe cite insanity or madness. It is noticeable that we speak today not of "paradoxers," but of "cranks," a word that comes fully loaded with all the negative baggage De Morgan cautioned against. De Morgan specifically defended his discoverers against the charge of madness, and on this issue also he turned the table on the accusers: "It is a weakness of the orthodox follower of any received system to impute insanity to the solitary dissident," he wrote in the *Budget*. Madness is all too easy a charge, and besides, De Morgan said, the establishment itself has always been rife with lunacy. "If misconceptions, acted on with too-much self-opinion, be sufficient evidence of madness it would be a curious inquiry what is the least-percentage of the reigning school which has been insane at any one time."

As it turns out, De Morgan intended to write a companion volume to the *Budget* focusing on paradoxers within the scientific establishment. His widow, Sophia, described this unrealized project as "a second part, in which the contradictions and inconsistencies of orthodox learning would have been subjected to the same scrutiny and castigation as heterodox ignorance had already received." The *Budget* gives us a taste of what such a volume might have looked like in a few examples De Morgan cites from the highest echelons of science.

"The mathematical and philosophical works of the Right Rev. John Wilkins . . . In two volumes, London, 1802, 8vo," reads one entry in the *Budget*.

John Wilkins (1614–1672) was a master of Trinity College, Cambridge, and one of the founding figures of the Royal Society. In 1638, Wilkins published a book of scientific speculations that included his thoughts on the possibility of flying to the moon, a feat he proposed was accomplished by wild geese. "From the sublime to the ridiculous is but a step," De Morgan scoffed. "Which is the sublime, and which the ridiculous, every one must settle for himself." De Morgan gives us a few other examples of what he considered lunacy within the establishment, including one from Newton's contemporary Christiaan Huygens, one of the most brilliant physicists of the seventeenth century. In his book *Cosmotheoros* (1698), Huygens had speculated that there should be plants and animals living on other planets, a proposition that De Morgan declared to have no scientific foundation. As De Morgan viewed it, interplanetary travel and extraterrestrial life were absurd ideas that had no place in the minds of professional scientists and that served only to show "how very near pure speculation is to fable."

Who is sublime and who ridiculous? That is a question we must each settle for ourselves.

A Budget of Paradoxes gives ample proof that discoverers "are frequently undisciplined intellects." The term is Ernest Nagel's, and following De Morgan, Nagel insists in his introduction to the *Budget* on treating paradoxers with respect. These are "often men," he wrote, "with unusual mental powers, and not seldom with a highly developed sense for logical cogency." In short, discoverers may be ignorant, but they are not stupid. They may be uneducated, but they are not dumb. The transition from the term *paradoxers* to *cranks* over the past century marks a shift along the axis of derision that downgrades the dissident scientist to the level of a fool. But as Nagel rightly notes, many outsiders *do* have a highly developed sense of "logical cogency." Often as not they are intelligent people passionately engaged with ideas, who have thought through at length what they see as significant lapses and inconsistencies in mainstream science. Not infrequently, the holes they see *do* relate to unsolved questions, for there remains a great deal in physics that is unknown, and almost every week now insiders themselves offer up radical new ideas about gravity, space, and time.

Yet if it is not madness, or a desire to deceive, that drives the outsider physicist, to what forces may we attribute the outpouring of theorizing that De Morgan documented in the nineteenth century and which, judging by my own collection, has been going on ever since? If Jim Carter were an isolated case, we might put his theories down to delusion or hubris or a lack of educational opportunity, but *A Budget of Paradoxes* and the vast empire of the Natural Philosophy Alliance serve to suggest that no such easy answers will do. In the mid–nineteenth century, it probably wasn't possible to theorize about the theorizers. A century and a half later, I believe there are things we might say.

In order to have a clear class of *paradoxers* there must, by definition, exist a well-accepted *orthodoxy*. The mid–nineteenth century was the moment in history when science in some sense crystallized. Before then, the very idea of "science" was still hotly debated. As De Morgan's views on extraterrestrials demonstrate, what was orthodox and what was heterodox depended on whom you asked. The word *scientist* was indeed coined in 1834, specifically to name the emerging group of science *professionals.* Before then, "scientists" had been "natural philosophers," the old-fashioned term to which the NPA's members now wish to return. As professionalism took hold in science in the latter part of the nineteenth century, the encompassing grace of that gentler term fractured into ever more specialized shards, giving rise to crisp new fields such as "physics," "chemistry," "biology," "geology," and so on. Along with that came separate university departments, separate professional societies, separate working methods, and, increasingly, an inability by those in one discipline to speak to those in another. In this landscape of ever more finely parsed levels of expertise, the "ordinary" citizen became ever more distant a spectator. One might well read popular books and attend the public lectures, but words could take you only so far. To really understand what was going on in physics, one had to learn the math. Even if a citizen wanted to keep up, there was no way he or she could. Shut out of the academic sphere and unable to comprehend the language physicists now spoke, the nonprofessional had little choice but to give up on understanding or go it alone.

The increasing professionalism of science is in some deep sense the reason I am positing for why the "flies" swarm—it constitutes the sociological sense. Yet as Darwin noted around the time the *Budget* was being compiled, individuals are never just products of "group" dynamics. We see the world through our own eyes, not through those of our species. And so we may refine our question: What is it that an individual human gains from having his or her own theory of physics? How does this assist his or her experience of the world? And what are the prices paid for this evidently risky behavior? Darwin raised pigeons to gain insight into the grand march of biology; so, too, can we look to a specific case for insight about the process of physics on the fringe. While the forces of professionalization produced the new class of physics experts and thereby created the conditions for the emergence of a paradoxical *class*, in order to understand a peacock, we must consider its particular case.

Part Two

JIM'S WORLD

Indeed, I can think of no one who has been anything but detrimental to the creation of this work.

—Jim Carter,
The Circlon Atom

Chapter Four

THERE'S DIGGERS, AND THERE'S EVERYONE ELSE

ON OCTOBER 4, 1957, the space age began. The world looked on as the first orbiting satellite, *Sputnik 1*, whizzed through the night sky, extending the range of human craft to the celestial realm. Launched from Site No. 1 at the Tyuratam range in Kazakh—now also called the Baikonur Cosmodrome—*Sputnik* traveled at eighteen thousand miles per hour and took just ninety-six minutes to complete an orbit around the earth. Like an electronic angel singing from on high, it emitted radio signals that were monitored by amateur operators across the planet. The signals continued for twenty-two days until the transmitter batteries expired, and two months later, on January 4, 1958, the satellite burned up as it fell from orbit and reentered the atmosphere. In its brief life, *Sputnik* traveled thirty-seven million miles.

Sputnik electrified the U.S. scientific and engineering communities. Since the end of World War II, Americans had believed themselves to be at the forefront of technological innovation, yet the Russians had beaten us to space. While the United States had been focusing on airplanes—Chuck Yeager became the first person to break the sound barrier in 1947 in the Bell X-1, and in 1953, Scott Crossfield was the first to fly at Mach 2—the Soviet Union had been quietly pouring its technical know-how into mastering the "high ground" of space. The success of *Sputnik* sent waves of alarm across the U.S. political landscape, for we were already deeply immersed in the cold war mentality. Three years later, the Soviets achieved something even more impressive: On April 12, 1961, they sent Yuri Gagarin into space, making him the first human to travel outside the earth's atmosphere. As Tom Wolfe documented in *The Right Stuff*, these twin Soviet triumphs created a crisis mentality among U.S. military officials and precipitated the decision to begin our own space program. We would soon be on our way to the moon. The unexpected genius of the Soviets precipitated another discussion as well—about science education—for how could America retain its preeminence in technological fields and keep up with the Russians if its children weren't getting proper schooling about scientific basics? If the new political reality of the space race created far-reaching consequences for the spectrum of American science, one of the most productive was that during the 1960s a lot more resources were channeled into science teaching.

Born in 1944, Jim Carter was just old enough to miss this trend. *Sputnik 1* was launched when Jim was thirteen years old, and it was not until he had finished high school that the new focus on science pedagogy would begin to be felt in classrooms across the nation. It is more than likely that the physics master at White River High in the late 1950s was not a first-rate teacher. King County with its fir forests and its small family farms was a long way from major centers of learning, and country schools often don't attract or keep great teaching staff. White River

High, which still stands today on a sprawling eighty-four-acre piece of land in Buckley, had a student population drawn from the surrounding areas. Many of the children were likely destined to stay on the land, and science teaching probably wasn't at the top of the school's priorities. Jim himself is wont to remark that at school he learned to read and not much else, and it is interesting to ponder what path he might have taken had he had access to a first-rate science education. Einstein once said that in science the hardest part is the questions; what makes for a good scientist in his view was not so much the ability to come up with answers as the perception it takes to ask penetrating questions. In high school, Jim began to ask questions that were beyond the range of his physics teacher's ability, and the answers he received often seemed to contradict what he was learning on the farm.

As a farm boy, Jim was familiar with guns, and one day his teacher used Newton's laws of motion to explain what happened when a rifle is fired. For every action there is an equal and opposite reaction, and according to the teacher, the kinetic energy of the bullet is equal to the recoiling energy of the gun. Jim objected. He had used high-powered rifles to shoot targets in the woods, and he had seen the impact bullets made in trees. He had felt the effect of the rifle's recoil on his shoulder and he did not believe its energy was anywhere near that of the bullet. He argued the point with his teacher, who stuck to his own guns. "I tried to make it clear to him that he was wrong," Jim recalls, "and he made it clear to me that I had to go to the principal's office." Jim pressed his case to the principal. The man listened patiently, amused by his student's passion, but he did not seem to know who was right or wrong, and in any case he was unprepared to adjudicate on a scientific matter. As Jim tells the story, "His solution to the problem was that it didn't really matter who was right. If I had strong oppositions to my teacher, it would be best to keep them to myself."

Some months later, the teacher introduced the mathematical principles behind Newton's laws, and he now explained the difference between *energy* and *momentum*. It is momentum, *not* energy, that is equally distributed between a gun and a bullet when a gun is fired. Jim felt vindicated, but the incident exacerbated his already nascent distrust of authority, and more arguments ensued. Often, Jim acknowledges, he was wrong, but occasionally he would "catch the tiger by the tail," and gradually he started to believe that "an authoritative opinion might not be as good as an ignorant but intuitive opinion."

Jim recounts the story of his high school years in a book he has begun to write called *Along a Twisted Trail of Truth: The Autobiography of a Science Crackpot*. He has always had a sense of humor, and "crackpot," he says, is a powerful word. Reading this lively little volume, one can't help feeling a bit sorry for the teacher. The man had probably taken only a few college courses in physics himself—to confuse energy and momentum is a basic mistake—so when Jim brought up the subject of rockets, it is hard to see how the fellow stood a chance. Again Newton's law of action and reaction was the point under discussion. The teacher said that for any object to move, it must push against another object: "A car pushes on the road with its tires, a boat pushes on the water with its propellers, and a jet plane pushes on the air with its engines," he told the students. Jim understood about tires and propellers, but was skeptical about the larger claim and asked the teacher how a rocket traveling in outer space could be pushing against *anything*? Wasn't space supposed to be empty? With the nation now transfixed by the dream of going to the moon, the question was more than rhetorical. Rockets traveling through space weren't fantasies anymore; we were in the business of building them, and understanding how they worked was a critical issue if we were going to succeed beyond our planet's atmosphere.

Addressing Jim's question, the teacher replied that space was filled with *ether*, an invisible substance pervading the universe. Like a boat pushing against water, he said, the rocket would be pushing against this ethereal substance. This was the first time Jim had heard about an ether, and even as a teenager he "didn't like the idea." As he points out in his book, "A rocket doesn't need anything to push on in order to accelerate. What accelerates a rocket is the transfer of momentum between the rocket and its exhaust." This was Space Science 101.

Outside class, Jim would read encyclopedias in the library, and most of what he learned in school he taught himself. Seeing his teacher be wrong, and more important seeing his teacher insist that wrong was right, encouraged his skepticism about authority and particularly about instructors. He had always been prone to pursue his own path, and his experience of high school science cemented in his mind a belief that if he wanted to understand how the world worked, he was better off trying to figure things out for himself.

After he finished high school Jim did odd jobs for local farms and sawmills. He needed time to spread his wings before considering further education, and in the early 1960s the pressure to go to college was a good deal less than it is today. Living at home in Buckley was an inexpensive option, and as his relationship with his parents had never been problematic, he was happy to help out on the farm in return for room and board. While he was chopping wood and mending fences, he could think about physics, letting his mind wonder about the mysteries of motion.

In Jim's accounts of his early life, there is virtually no mention of the extraordinary cultural context in which he was beginning his adult years: The Vietnam War was raging; folk music was beginning to colonize the airwaves; in the nightclubs of Greenwich Village, Joan Baez and Bob Dylan were singing songs of protest. To be sure, Buckley was a long way from Manhattan, but it was impossible to be a young man in America and not feel something of the era's pull. Bob Dylan, one of the few people Jim has been heard to admire, was exhorting his generation to cut loose from institutional ties, to let go, be free, and experience the world for itself. At the end of 1962, a few months after he completed school, Jim decided to hitchhike across America, thumbing his way with strangers and doing whatever work he could pick up along the way. With a few dollars in his pocket, he bade farewell to Buckley and set off for wherever the road might take him; he was Dylan's ramblin' man.

On that road trip across America, Jim slept in the backs of trucks, loaded corn, dug trenches, helped deliver produce to warehouses, and encountered his first McDonald's. He found the McDonald's in Milwaukee when a group of teenagers dropped him off at its door. Driving down from Minnesota, he had been daydreaming about magnets, and when he saw the huge yellow hoops outside the franchise the thought popped into his mind that they were "like big circular magnetic fields attracting customers." He had spent days living on corncobs, and in one sitting he devoured six hamburgers. Along with initiating a lifelong love of travel, this road trip gave Jim his first opportunity to engage with a professional physicist, for just as he was leaving McDonald's he picked up a ride in a brand-new Cadillac Eldorado driven by a retired professor of physics from Wisconsin. The professor was headed for Denver, and on the long drive south Jim "pumped" his driver with questions. Eventually the pair of them got to talking about gravity. It turned out the professor had his own ideas on the subject, something to do with an ether. Jim wasn't interested in ether but what really caught his attention was the professor's admission to being a "dissident." By the time he got home to Buckley, he had found his vocation; he knew now that understanding the universe was his principal goal in life.

In January 1963, Jim enrolled at Pacific Lutheran University (PLU). It was the second semester of the academic year, and to his disappointment there weren't any freshman physics courses offered midyear. Instead he took psychology, public speaking, and a mandatory religion class. He was also required to attend chapel every day. By now he knew better than to argue with professors, and he spent his energies debating the nature of reality with his fellow students. Both the bright young things, or "Jump Jumps," as they were called in his dorm, and the biblical literalists, or "Lute Fruits," offered equal fodder for his quizzical mind. By spring break it was clear that the university probably wasn't the place for him to be; the kind of work he wanted to do "on an original theory could never be done in a university setting," he concluded. And so it was that halfway through his only semester of tertiary education, Jim Carter put aside his books and went off to search for a meteorite. It may not have been a canonical step in the life of an aspiring scientist, yet neither could it be termed an entirely illogical move. If Jim was abandoning a formal framework for the development of his ideas, it was only because he had a far bigger framework in mind.

The prize that lured him was the so-called Port Orford meteorite, a chunk of cosmic debris that had originally been discovered in the mountains of Oregon in the mid–nineteenth century. Its location had been lost when its discoverer died suddenly without telling anyone its whereabouts. Reputedly, this piece of space rock was worth several million dollars: "A holy grail of forbidden wealth," is how one enthusiast described it. As with one of those Spanish galleons off the Florida coast, men *knew* it was there, if only they could find it again. The Port Orford meteorite was supposed to be one of the rarest kinds. As Jim would later write, "Careful analysis of this pallasite may reveal to scientists information about outer space and the creation of the universe which may never be known if the meteorite was not recovered." Meteorites provide evidence of the processes by which planets are formed and pallasites are particularly rich specimens for illuminating these mechanisms. Looked at from this angle, skipping out of school to find a missing meteorite would be a realization of the very ideals the professors were espousing. With humanity poised to go into space, the more we understood about pallasites the better off we would be. Thus began Jim's career in physics on the fringe and a lifetime spent pursuing dreams in wild and unruly places.

From the 1850s, the Port Orford specimen had been a subject of intense speculation in meteorite-hunter circles. Dr. John Evans, a physician-turned-geologist working for the U.S. Department of the Interior, had supposedly discovered it in 1856 while surveying the territory of Oregon. Dr. Evans had come to this task with a reputation as a fossil hunter and had brought with him a keen eye for geologic anomalies, so when he spotted an enormous metallic-looking boulder near the Rogue River, his curiosity was piqued. According to Evans's journal, only the tip of the meteorite protruded above the ground, but this bit alone was five feet high. With immense effort he managed to break off a piece that he brought back home for chemical analysis, which revealed it to be a pallasite. Common meteorites are composed of either iron or stone; pallasites, a mixture of the two, are believed to originate in the largest asteroids that, like our earth, have a liquid core. They are effectively protoplanets. Such meteorites may also contain crystals of olivine, a fact that seems to have inspired the makers of *Superman Returns*. In the film, the piece of kryptonite that Lex Luthor steals is a glittering green crystal said to have come from a pallasite. For planetary astronomers, they are pure gold. Evans estimated that the total weight of his find was twenty-two thousand pounds, making it by far the largest specimen on record.

News of this discovery generated excitement in astronomical circles, and Evans worked hard to assemble an expedition to retrieve it. Congress was on the verge of appropriating funds when on April 12, 1861, the Civil War broke out. The next day Evans died of pneumonia, taking knowledge of the pallasite's location into the next world. In the fullness of time, Dr. Evans's journal was acquired by the Smithsonian, and in 1929 the institution sent its curator of mineralogy to retrace his steps. Again in 1939 another curator was dispatched, but neither mission yielded many clues and the journal itself remained maddeningly short on detail. Treasure seekers' interest was sparked in 1937 when a Sunday feature in the Portland *Oregonian* estimated its value at $100 a pound, or more than $2 million total, a staggering amount of money at a time when the nation was still mired in the Great Depression. In 1940 Myron Kilgore from the Society for the Recovery of the Lost Port Orford Meteorite led yet another search party, and in 1950 James Karle, an astronomy teacher at Lewis & Clark College, spearheaded a further effort. In 1963, when Jim hitched his way to Portland, the purpose of his journey was to meet Professor Karle in person.

Sadly, Dr. Karle could offer no illumination as to the meteorite's whereabouts, and Jim returned to PLU to finish off the semester. But his mind was made up, and as soon as school was over he ditched academic life for good and took off in his Chevy to explore the terrain on his own. He would recount his exploits in a series of articles for the local *News-Banner*, a boys'-own version of Indiana Jones complete with treasures and dangers and a lost city to boot. What Jim lacked in funds he made up for in supplies, and the Chevy was piled high with camping equipment, climbing gear, carpentry tools, and a gold pan. As he planned to be in the mountains for several weeks at a time, he also took along "an enormous supply of brown rice, dried soup and dehydrated fruit." Rounding out the load were the notes he had been developing about a new theory of gravity that was evolving in his mind. Unlike his teacher, he was convinced he could explain the forces of the universe without resorting to an ether.

Most previous searches for the Port Orford meteorite had focused on a bald-topped peak known as Iron Mountain, but Jim determined Barklow Mountain as a more likely spot. His hunch was confirmed by a wizened prospector who greeted him at the end of a trail on his first afternoon. His newfound friend, Harry, had insider knowledge of the fabled object and claimed to have actually seen it. Later that evening as the two men sat over a simple dinner in Harry's hut, Harry regaled Jim with stories about his youth as a gold miner back at the start of the century. In 1911, Harry told Jim, he had been hunting for elk one day when his eye had been caught by a large, strange-looking rock sticking out of the hill. Round and shiny and "different to any rock" he had seen before, he had made a mental note of it. His mining partner, McCurdy, confirmed that he too had seen the boulder. "Far's I can make out," McCurdy opined, "it's either a large jade boulder or one of them meters."

In 1911, few people placed any value on "space rocks," and Harry gave the "meter" little thought until the appearance of the *Oregonian* article twenty-five years later. He could no longer recall its whereabouts, but he told Jim he was certain it was on one of Barklow's ridges. Lying in Harry's hut that night pondering the future of physics, Jim began to see that science might offer a solution to the meteorite's location. Pallasites are made principally of iron, so during summer the metal would heat up. Jim reasoned that it should be substantially hotter than the surrounding earth, and it occurred to him that if he took a series of infrared photos of the mountain, a huge lump of iron like this might show up as a hot spot on the film. The next day he purchased some film and set about taking pictures. His aim was to assemble the individual shots into a photomontage, a technique that the British painter David Hockney would later make famous as a new form of art. Unfortunately, when the images came back from the developer, it wasn't possible to see the level of detail required, leading Jim to conclude that he would have to give up his quest.

Yet a car full of brown rice was not something to abandon lightly, and Jim soon saw that while one door had been closed, another was about to open. As he listened to tales from Harry about his early years as a gold miner, Jim found himself gripped by a different kind of prize. Here is how he describes the transformation in his *News-Banner* report:

> In the eerie half-light of dusk, I could almost see the red-shirted miners as they slowly trudged their way up to their cabins after a hard day's work at the diggings. I began to wonder what this country must have been like when old "Course Gold" Johnson first discovered gold here over a century ago.

According to Harry, the old-timers had taken more than $5 million from Johnson Creek alone, and there was "plenty more in these hills." While Harry's words sank in, Jim contemplated the moon rising into the heavens like a freshly minted coin, and in the suggestive bath of its light he experienced an urge as deep as any encoded in the human brain. "I could feel the lure of gold take over me as it had done to countless thousands of others since the dawn of time," he confided to his readers. The Chevy full of rice and beans would not go waste. The search for an iron meteorite would be transmuted into a search for gold.

Had Jim been born in another era, he might have found a niche as one of the underlings on an expedition to the New World or the South Pole. Explorers from Hernán Cortés to Ernest Shackleton were always on the lookout for resourceful individuals like James Carter, but few other eras would have enabled such expression of his own exploratory zeal. In the early 1960s, the world was an open book to young persons of adventurous leanings, and while some found bliss in the monasteries of Tibet or the drug trails of Afghanistan, Jim found his nirvana in the mountains of Oregon. He had come in search of a messenger from the stars; instead he would spend the following summer digging into the bowels of the earth. In the process he would work harder than on a prison chain gang. He would rise at dawn, break rocks all day, and go to sleep exhausted. If it hadn't been a free choice, it would have been some people's idea of hell. But the brutal regimen Jim lived during the summer of 1964 was his choice, and by his own assessment, the three months he spent at the "Blue Hole" remain among the seminal experiences of his life.

Jim had learned about the Blue Hole from Harry. In rivers that run with gold, miners know that the best places to look are the deepest: Gold is nineteen times denser than water and twice as dense as iron, so it tends to sink to the bottom and accumulate in pits gouged out by turbulence. Experienced miners search for such holes, but in regions as well picked over as the Rogue River, few depressions would not have already been "bottomed." Miners, like meteorite hunters, are however an optimistic crew, and in the landscape of prospectors' desire there are always new hordes of gold waiting to be discovered. In short, the Blue Hole was the miner's version of the Port Orford meteorite, a legend that had taunted treasure hunters for a century. Though many had tried, none had succeeded in bottoming it.

In contrast with the meteorite, the location of the Blue Hole was known precisely. As Jim described the place in his *News-Banner* report, it looked less like the scene of an assault than a private piece of paradise. Located deep inside a thick forest, it can be reached via a logging road, then a half-mile trek through a canyon. It wasn't the kind of place you'd stumble on if you didn't know it was there, and even if you knew, it was a lot of effort to reach. At the end of the canyon, rock walls opened out suddenly to reveal a spectacular hidden glen of Port Orford cedars. Dotted through the glen were sword ferns, and a carpet of moss swooshed across the floor. A stream ran through the clearing, and behind it rose a cliff framing this miniature Eden. Abutting the cliff was a wide granite shelf that dropped away sharply on one side. The stream flowed across the flanks of the shelf and over its edge in a cascade of spray. Over aeons of time, the water had carved a huge, almost perfectly circular hole in the edge of the shelf some fifty feet across and fifteen feet deep. The pool at the bottom of this hole was surrounded for most of its circumference by the wall of the rock, and down its flanks tumbled a waterfall. The only thing amiss, Jim remarked in the *News-Banner*, was that "the person who named it must have been color blind, for the predominant color was green not blue." Whatever the hue of the water, the tone that mattered was the color of the gold imagined to glimmer at its bottom. "As I gazed into this wonder of nature I could feel it challenging me as it had challenged so many others," Jim wrote. "I knew then and there that I would conquer this hole if it were the last thing I did."

In order to reach whatever gold might be lying underneath, the whole thing would have to be cleared of gravel. Given the pool's scale it was a monumental task, and Jim later estimated he would clear eight hundred tons of rock. By his reckoning, the key to the operation lay in getting the right equipment, so he set about amassing supplies. At the top of his "equipment list" was "a partner," and he knew just the right man for the job. "I had seen him display amazing endurance when we climbed Mt. St. Helens," Carter reported in the *News-Banner*. "I knew he would have the determination to match his enthusiasm." In Steve Loftness, a chemistry student he had met at PLU, he recognized an equal aptitude for punishment.

Jim figured that the way forward was to build a dredge to suck the gravel out of the pool, then run the gravel through a sluice box to separate the gold. One of them would tend the dredge while the other used diving gear to vacuum the bottom of the pool with a suction hose. He had read about such setups and he signed up at a local technical college so he could construct his hydraulic dream. To earn the money to fund this adventure, he worked at a sawmill, lumberjacking by day and machining in the college shop by night. After work, he and Steve would get together to talk about science.

On June 10, 1964, Jim, Steve, and Jim's brother, John, took off from Buckley, three young men with gold dust in their eyes, leaving behind a world that was currently sending some of their friends to an incomprehensible war. For the next three months, the Blue Hole glen would be their workplace and their home. "Home," if you could call it that, was a shack constructed from poles covered in plastic sheeting. When it rained they got soaked, and nothing seemed to keep out the cold. They dragged in a wood-burning stove to cook on, but it filled the shack with smoke and left them no less chilled. To call the place a "hovel" was a term of affection in Jim's eyes and the preferred wording in his *News-Banner* reports.

The descent into the Blue Hole was inaccessible by car, so everything they needed had to be carried in on their backs.

By day they heaved equipment, by night they fantasized about how to spend the gold and engaged in long discussions about the foundations of science. Jim had been working alone on his gravity theory, but in talking to Steve, he found himself drawn to another question: What is the structure of matter? Under Steve's tutelage, Jim started to learn about the periodic table and about the properties of each atomic element, the inertness of helium, the bonding capacity of carbon, the reactivity of sodium, and so on. Every atomic element has unique properties. What is it that causes these qualities? What is it about carbon, for instance, that makes it behave so differently from nitrogen, which sits directly next to it in the periodic table? There must be something about the structure of each atom that produces its specific properties. Talking to Steve at the Blue Hole, Jim began to feel that here was a question worthy of his attention.

The idea had been to spend the summer dredging gravel out of the Blue Hole, and if they didn't finish by the end of summer, they would return next year to complete the job. But just as they were about to start, local miners told them that winter flooding would refill the whole pool with gravel. If they hoped to bottom it, they would have to do so in a single season. An already difficult task began to look impossible, and more prudent men might have paused to wonder if the effort would be worth it. Jim, however, was set on a goal, and in a flash he saw the solution: They would stop the flow of the stream, diverting the waterfall and damming the creek. They would build a flume over the pool, to carry the water away in a kind of aqueduct, then instead of using the dredge they would clear away the gravel using shovels.

This improbable scenario was rendered possible by a discovery they had made while hiking in the woods. Nearby they had come across an abandoned mining town, called Galena, that had burned to the ground in 1911, the same year Harry had seen the "meter." The town's major asset had been a stamp mill where ore from local gold mines was ferried in for processing. Scattered throughout the site were the remains of the old mining operation, including a huge old flume that had carried water to power a giant waterwheel. These two massive objects had been brought in by the old-time miners using the same trails Jim and his friends were now using. These objects had been hauled in by men undaunted by the prospect of backbreaking labor. With a sense of awe, Jim described the Galena site in his *News-Banner* report, marveling that "everywhere were the rusting and decaying remnants of what man could do if he had enough determination." In Jim's lexicon, "man" is always a personal noun: If the old-timers could do it, then why couldn't *he*? Why couldn't he and Steve and John, three able-bodied laborers with wills as steely as any pioneers repeat the task? They could pull down the mill's smokestack and drag it back to their own camp, where it could serve as *their* flume. The spirit of Galena could be resurrected in their own mining complex.

It is hard to imagine the reaction that might have greeted this suggestion as Jim spelled out his plan to Steve and John. Alone in the woods, they had no carts, or mules, or even a radio in case of an emergency. Some of the trails were pitched along the edges of cliffs. Never one to dwell on negatives, Jim admits in the *News-Banner* that his companions' initial response was "skeptical." Unfortunately, their thoughts are not included in his report, which moves on swiftly to "wholehearted agreement." So there they were the next afternoon, dragging the iron smokestack through the forest and narrowly missing several close shaves on the cliff.

The smokestack formed the basis of a flume that Jim designed to divert the flow of water from the falls, and they rigged it up to the surrounding trees with cables scavenged from the Galena site. The dam offered its own challenges. Jim had decided it should be built from mud and rocks held together with moss. When this beaverish fantasy failed, they reluctantly built a formwork and poured cement. After two weeks of backbreaking labor, the flume, the dam, and the dredge were in place and they were finally ready to begin clearing rock.

Word was getting around about the kids going feral, and visitors began to trickle in: Local miners, a couple of school friends, and Jim's parents made the trek. The scene resembled something out of *Lord of the Flies*. No one who came could believe the state of their living quarters and even Jim acknowledged that the charms of the hovel might be hard to see.

The thing that most transfixed visitors was the sight of his sleeping bag, an old World War II "mummy bag." "It wasn't so striking in itself, but it was responsible for the thing that was." Through a rip in its side, the bag was leaking feathers that encrusted every surface of the place with a ratty pouffing of down, giving the hovel the appearance of a rancid nest. The smell of stale peanut butter wafted through the air, along with the aroma of rotting mouse, for in addition to feathers, the shack was overrun with rodents. Jim had tried all-out warfare with an economy-sized tub of rat poison, but it took the mice just two nights to polish off the can and, as he noted admiringly in his *News-Banner* report, "it didn't slow them down a bit."

Visitors were fed the same rations the boys were chowing down every day: for breakfast, pancakes one morning, oatmeal the next; for lunch, hardtack with peanut butter and jelly; for dinner, boiled brown rice one night, fried potatoes the next. In a gesture of self-preservation, people took to bringing in their own food, and in the *News-Banner* Jim recalls a feast of fried chicken and apple pie his parents provided out the back of their station wagon. Fueled by a diet that would leave many of us too listless to get off the couch, Jim and Steve set about clearing the Blue Hole of gravel. At this point, John threw in the towel and headed back to Enumclaw "to mow hay."

Jim's father offered to build them a conveyor belt to move the bigger rocks, which sounded like a splendid idea until its gears broke down under the stresses of the rugged terrain. Trouble compounded a few days later when an application of dynamite blew the flume to smithereens. At this nadir of their competence, it became clear that no kind of mechanical help was going to be of any use, and in a letter Jim acknowledged that "the only way to get to the bottom was to get down and slave." Abandoning all hope of assistance, he and Steve ditched the conveyor belt, rebuilt the flume, and relinquished themselves to the path of manual labor. They adopted a system they would keep up for the next two months—at the bottom of the Blue Hole, one of them would shovel gravel into a five-gallon bucket that the other would carry up a log stairway and dump into the sluice box. Thousands of pounds of rock would be lugged up this way, all as a personal choice by people who were supposedly "dropping out."

Many years later at the Green River Gorge, I asked Jim to explain to me the charms of this situation. For most people, digging is a form of punishment—it's the archetypal penal task—for Jim, it seems close to a sacred activity. "When it comes to digging," he told me, "there's diggers and there's everyone else." Far from being a means to an end, Jim sees digging as its *own* reward.

What makes for a good digger? I inquired.

"Well, it's someone who *likes* to dig, someone who appreciates the *value* of digging."

And what exactly is the value of digging?

"It's kind of like a meditation. When you get through the end of the day you can see what you've done and see that you've made progress."

At the Green River Gorge, Jim would put his love of digging into practice in what may be the most mysterious project of his life. Deep in the cliff face at the bottom of the gorge, he has hand-chiseled a cave. You enter it by tunnel, whose entrance is hidden in the brush so that no one looking from the outside would have any idea it was there. The only way to get in is to crawl on your belly through the mud along a tiny hidden passageway. Inside, the cave itself is the size of a modest-scale room gouged out of the sandstone. Around the walls swirl layers of colored stone, red and orange and gray. The reason you can see any of this is that Jim has installed a power line and a naked light bulb hangs from the center of the ceiling. In addition, there is hot and cold running water and a telephone. The whole exercise took about eighteen months to complete, with Jim working on his own using a handheld jackhammer. Not only did he chisel the space by himself, but all the rock displaced he carried out on his own.

When you ask Jim why he went to all this trouble, he shrugs and looks inscrutable. "I just liked the idea of a cave," he says. Letting his imagination wander, he goes on to speculate about an underground network of interlinking caves and passages with windows looking onto the gorge.

Back at the Blue Hole, "progress" was definitely being made. It could be measured by the yard. In another letter home, Jim wrote:

> In one place we are nearly down to fifteen feet and the bedrock is still dropping straight down. We are probably the most famous miners in all the Rogue River Mountains. Old-timers come from miles around to see the kids who are doing what they all said was impossible to do. Everyone says we will strike it rich for sure. No one who comes here can believe that two guys could dig such a gigantic hole.

By the start of September, the hole was so deep that when Jim and Steve finished work each night they had to climb up two sets of ropes to get back to camp. Physically exhausted, they would spend their evenings thinking about the periodic table and the problem of atomic structure. What is it that causes the pattern of rows and columns that all the atoms obey?

Despite the fact that they had taken fifteen feet of gravel out of the Blue Hole, they were still finding old nails and bits of tool among the debris, evidence that others had been there before them and a likely sign that the gold was already gone. The most disheartening evidence of previous attempts appeared after more than eighty days of digging, when they discovered a couple of poles protruding through the rubble. Was all their work in vain?

Several more days of digging revealed that the poles were pry bars wedged under a boulder. Here at last was a happy sign, for if the boulder was still there, then no one had dug beneath it. Two days later, Steve picked up a nugget "the size of a pencil eraser," and as they fell asleep that evening, they had every reason to believe their fortunes were made. Yet nature had in store a final travail. Halfway through the night they were awakened by rain. As the storm pelted down on the hovel, Jim and Steve imagined the Blue Hole filling with gravel, and above the din they speculated about their summer literally going down the drain. When finally the tempest stopped, they tramped outside to inspect the damage, peering into the Blue Hole, which now, "took on the appearance of an extinct volcano." The bottom was engulfed in darkness. Nothing would be known until morning.

Moving toward its cliff-hanging conclusion, Jim's *News-Banner* account slips into an almost beatific calm. Here is how he and Steve begin their next day:

> Soon the moonlight gave way to sunlight and the song of the bubbling creek was rivaled by the songs of numerous birds. Steve started a fire and soon the smell of fresh coffee over-powered the smell of wet canvas. A doe and her two

fawns came down to the creek to drink, still wet from the downpour, which now seemed a bad dream.

Dreams were all they would take away, for when the last of the gravel was cleared the Blue Hole proved to be barren. After three months of punishing labor, all they had to show was a handful of nuggets, a few hundred dollars for a summer of their lives. Steve summed up the situation in his diary, the one comment we have from Jim's heroically silent partner: "Cleaned the hole today but there was no gold. The biggest mining bust in the whole history of mining." Piecing together the mystery, Jim saw that the problem lay in the Blue Hole's perfection: When a flood had swept through the region three or four hundred years earlier, it must have taken out the gold along with the gravel, for the bottom of the pit was "smooth as a baby's bottom." Had its floor been less perfect, Jim speculated, it could easily have held a fortune. To many men that might have come as a disappointment, yet if there was any doubt in Jim's mind that the time had been well spent, there is no evidence in the *News-Banner.* On the contrary, his assessment of the story's ending was positively upbeat:

> At least we had the satisfaction of solving the hundred year mystery of the Blue Hole. No more will treasure hunters gaze at the beauty of this magnificent water pool and dream of finding gold at the bottom—for everyone now knows the answer.
>
> But the Meteorite—the Black Fortune of the Rogue River Mountains—is still an unsolved mystery and is the

prize so many are seeking. I know of one sure way to find it—but it will take quite a bit of money! Some day, if fortune smiles, I or some other lucky person will find it.

Ever optimistic, Jim sees the meteorite not as a lost or hopeless cause, but as a symbol of possibility, a prize that is out there waiting still for some fortunate soul.[1]

Chapter Five

THE FOUR SEXES

PACIFIC LUTHERAN UNIVERSITY might not have provided Jim with an education, but he did not come away empty-handed. It was there that he started dating a young social work student named Linda Vick, the woman with whom he would spend the rest of his life. It would be tempting to say that in this case also fate has smiled on Jim, for no one who knows the couple could deny that with Linda he has been fortunate in his bride. In photos from the 1960s she looks steadily into the camera, calm and smiling, her hazel eyes wide-set, her hair brown and long. She has always been a curvaceous woman in an earthy Italianate way, and today in her sixties she retains a youthful fitness, even though—and perhaps because—she now has four grandchildren to help care for. At the Green River Gorge, she tends a large vegetable patch, and as she leans over to pull weeds her back is supple, reflecting along with excellent genes a life spent vigorously with much of her time outdoors.

Linda had grown up in Los Angeles as the beloved and protected daughter of a Lockheed engineer; toward the end of her father's life she learned that he had worked on the Blackbird stealth plane. Her mother, a child of Italian Catholic immigrants, had converted to the Lutheran faith as a young woman, and it was she who decided on PLU as the appropriate college for her child. Mrs. Vick hoped that Linda might meet a minister at PLU and come back home with a nice churchgoing husband. If Jim didn't exactly fit the desired spousal model, he brought to the relationship a degree of self-awareness that few young men possess and with all his singularities he has proven to be both a steadfast and exciting husband.

Often people obsessed by physics are inattentive to their own and others' emotional needs, preferring the certainties of nature's laws to the chaos of their fellow human beings, and Jim can also appear this way. But in 1988 he published a pamphlet in which he outlined a theory of human relations. The coming into being of a man and a woman was Jim's theme here: How do we become who we are as psychosexual beings? And what exactly are we, anyway? In *The Four Sexes*, Jim proposed a theory of gender in which he posited the existence of four basic "sexual types." Though neither he nor Linda is mentioned in the booklet, within the framework his theory lays out the pair as constituting the perfect synchrony of what Jim calls the "yin woman" and the "yang man": two halves of a complementary whole who complete the space of one another's being.

Here we see in the outlines of this theory many of the qualities Jim would bring to bear on his physics, a mind engrossed with the totality of things and a preoccupation with origins. We see also the force of an engineering sensibility in which every element of a system has its *necessary* place in the structure of the whole. Just as yin and yang combine to form a richer unity, so we find in all Jim's theories a synthesizing of pairs: man and woman, mass and motion, space and structure, matter and antimatter. Above all, in *The Four Sexes* we observe the role of symmetry, a concept that would become central to all of Jim's ideas. To physicists, symmetry is one of the most important qualities in nature, and mainstream physics now holds that the fundamental nature of reality can be understood in terms of mathematically encoded symmetry relationships. Yet as Jim noted in *The Four Sexes*, the science of biology has been guilty of a major *asymmetry*: In the development of an embryo, it is the sperm alone that determines what sex we become. This is the imbalance that Jim sought to correct.

Biology tells us that our gender is determined by a sex chromosome that is passed on through sperm. While all eggs have an X chromosome, sperms have either an X or a Y; the former leading to a female offspring, the latter to a male. Mainstream biology leaves it at this, with the egg as the blank slate onto which the sperm projects its gendering stamp, but to Jim this inequity seemed illogical. If basic cell biology produces two different types of sperm, he argued, then surely "it stands to reason, and is even *required*," that it should also produce two different types of eggs. Mirroring sperm, Jim's theory posits two different "egg types," the "yin" and the "yang." According to him, where sperm account for our *physiological* gender, eggs determine our *psychological* gender, or "personality type." Balance is thus achieved and symmetry restored to the process of human becoming.

The upshot of this theory is that there are four rather than two essential "sex types," the yin-male and yang-male and the yin-female and yang-female. Each type must be understood as a separate category in the human taxonomy, for as Jim explained, the "four sexes differ from one another every bit as much in attitude and outlook on life as the two physical sexes differ in appearance and biology." Understanding these four types is essential for understanding human pair-bonding. According to Jim's theory, what matters most when a union forms between two consenting adults is not the physiology of the partners, but the psychological drama of the yin-yang interaction. In this way of seeing, the ideal partnership—regardless of the gender of the couple—consists of a yin together with a yang: "Quite simply, opposites attract and complement each other."

What tendencies and proclivities do we find in each of the two types? In Jim's words, the yang tends to be "outwardly directed," while the yin is "inwardly" focused. Decisive and self-motivated, yangs "crave control" over their environment and "at work will strive for a position of leadership such as foreman, supervisor or even president of the company." Yangs are "more comfortable acting on information received through the process of logic and reasoned thought," while yins act "through intuition" and are naturally inclined toward emotional engagement. They make excellent social workers and nurses and are likewise "the best doctors and teachers." With yangs' natural leadership potential, many of them seek professional autonomy, "finding satisfaction as taxi and truck drivers, or equipment operators, where they can exercise control over machinery." Not uncommonly, yangs strive to be self-employed and are forceful in initiating their own projects. Yins prefer to let others lead the way. Given the inherent symmetry of Jim's system, half of all men are yins and half of all women are yangs. Our incomplete knowledge of the fourfold panoply of gender has, in his view, straitjacketed our thinking, narrowing the range of expression for half the human population. In the final page of *The*

Four Sexes, Jim notes that there are "many experiments that could be done" to prove or disprove the theory. In truth, that is hardly necessary: Of the people to whom he has shown these ideas, almost all can "accurately identify themselves and others around them as either yin or yang."

There was no doubting Jim's type: "Self directed," relishing "control over machinery," a born innovator, he is every inch a yang, and in Linda he had found the embodiment of his yin ideal. After the couple moved to the gorge, Linda became a social worker for King County, specializing in cases involving badly neglected children. Her empathy and levelheadedness were qualities deeply valued by families and authorities alike, and she would often be called on to accompany children to court appearances. In the trailer park, Linda is the first shoulder to cry on and the arbiter of disputes. Around the campground, there are plenty of opportunities for perceived slights and Linda carries the burden of keeping the emotional peace among 120 people, some of them fairly damaged themselves. Although she is reluctant to intervene in her tenants' lives, when one of Jim's helpers was stabbed during a fight, it was to her that the community turned.

In the dynamics of Jim's theory, where the yin is the empathizer and the yang the activizer, Linda's warmth fills the void of what can sometimes appear as remoteness on Jim's part. "People often think Jim isn't interested in them," she has told me, "but it's just that he doesn't say very much." Having committed his thoughts on human psychology to print, Jim seems to feel no further need to elaborate on his ideas. The purpose of words in his world is not to enable social interaction, but to tame the chaos of unknowing and articulate a structure for reality.

As a yin cog to the driving wheel of Jim's yang, Linda has been happy to participate in a series of enterprises that many less "ideal" yins may not have found it in their power to embrace. In their playing out of the yin-yang dynamic, there have been many episodes in the Carters' life together that could be cited, but few would illustrate the case better than the events that occurred in 1971, around the time their son Paul was born. In the years since the Port Orford meteorite expedition, the couple had moved from Washington State to California. Linda wanted to be near her parents, so for several years in the late 1960s the two of them lived in L.A. Linda worked as a social worker while Jim tried his hand at a series of jobs, including being an officer for the LAPD. On the surface he had the right qualifications—he was competent with guns, he could cope calmly with difficult situations, he was young and energetic—but given his antiauthoritarian streak, it was always a mismatch, and a dose of reality quickly set in on both sides. Soon his entrepreneurial spirit began turning again to gold. This time the site of his imagining was the Feather River gold field in Northern California.

While Jim had recorded his adventures at the Blue Hole in the series of articles for the *News-Banner*, his adventures at Feather River would be documented in a series of 16 mm film reels that he shot on a Bolex camera in the summer of 1972. The footage is badly faded now and some parts have been erased by water damage, the colors bleached away. Shot without sound, the life on display in these silently flickering frames seems to belong to a bygone era. Thirty years after it was shot, I watched this footage with Jim. In the intervening years, it had sat unseen in the attic of the old emerald lodge, where rain leaking through the roof had been quietly leaching it away.

In the many hours of uncut film, no highways or electricity poles intrude upon the view. Nowhere is there a hint of "civilization." The scenery is massive, overwhelming, and wild. Vast cliffs tower on either side of a river that rushes around gigantic boulders. Today the Feather River Canyon has become a favored place for white-water rafting, but in the 1970s it was isolation incarnate. In the footage, Jim is accompanied by a small band of accomplices: Linda, his brother, John, and several other adventurous individuals are seen swimming and fishing and camping out. A dinged motorboat affords the group's only luxury, although mostly it is used for ferrying supplies for Jim's mining operation. In the film the young men's bodies are lean, their hair long, their beards straggly; the women, slender and strong, bask on rocks and float in pools, although they too must have done their fair share of hauling gear. An air of chaste robustness testifies that youthful entertainment need not come in pharmaceutical or electronic form.

As with most of Jim's endeavors, fun was packaged as a productive undertaking, and the footage records scenes of Jim and the other men dragging equipment over the rocks. Here too Jim had built a dredge and the men can be seen siphoning gravel from beneath gigantic boulders. One would be hard-pressed to imagine a less likely setting for a newborn baby, yet two months after Linda gave birth to their first child, Jim proposed that she and the baby join him in the canyon. As Linda tells the story, her parents were appalled. They had raised her in the bosom of suburban safety in the San Fernando Valley. During her childhood they had barely allowed her out of the yard. They were Italian. She was a girl. Her proper place was at home with a nice young man who had a steady job. At a family convocation around the packed car in which Jim prepared to whisk her away, her grandmother began shouting in Neapolitan: "You can't take a baby down there! What are you going to do without a washing machine?"

At the bottom of the Feather River Canyon, Linda didn't have a washing machine or any of the other accoutrements of modern family life. She did not need them and did not miss them, she says. She breast-fed Paul and was happy to wash his clothes in the river. Jim made a tepee from a parachute and built a cabin for them to sleep in. They stored their food in a cave. After her sheltered life in the suburbs, the adventure Jim offered seemed thrilling. "He had it all figured out," Linda recalls. "He knew how to survive in the woods. He had all the right supplies. He didn't have any sense that anything could go wrong that he couldn't deal with. I thought, I'll go along with this. It seemed pretty good to me."

Throughout 1972, the little family unit would shuttle between Feather River and the island of Catalina off the coast of Los Angeles, spending as much time as they could in the wild. Among the many reels of film there is a lovely piece of footage of baby Paul at two years old, floating in a rubber tire on the surface of a pool with the vast cliffs looming above him. It is a vision of childhood almost unimaginably remote from the world of iPods and computer games today, and watching his indolence, one finds it hard not to feel that perhaps we have lost something in our rush down the digital superhighway. In another scene, Linda and Paul are bathing in little circular pools carved out by the water in a massive rock shelf. Linda's hair streams around her bare shoulders while her son splashes nearby. It is as if some painter had framed a *Madonna and Child on Mars.* They could have been on another planet, and in many ways they were.

The footage documenting the Feather River adventure was itself the result of an improbable scheme. As at the Blue Hole, the California mining operation had proved less than a financial triumph. While they did find gold, it was never in sufficient quantity to pay for the expeditions. In the absence of treasure, Jim decided that if he couldn't make money *from* gold mining, then perhaps he could make money by making a film *about* gold mining. So he bought a Bolex camera and began shooting. He didn't have sound—"that was going to be added in later"—but he did have a couple of comely stars. He had placed an advertisement in an L.A. paper asking for volunteer extras, and in a city where every waitress hopes to be discovered à la Lana Turner, the slots were easily filled by two attractive women. Rounding out the party were the Carters' friends Henry and Cliff.

Sound wasn't the only thing missing. There wasn't any script, either. Jim insisted he had a plan for a narrative, and the details would be fleshed out as they went along. The spirit of the enterprise is captured in a lovely scene in which the little party is seen lounging on rocks in the middle of the river; a dog skips about in the spray, and nobody appears to have any agenda. Someone's backpack floats over the rapids, and nobody seems to care. That pack shot would prove proleptic. Three weeks into filming, Jim and Henry and Cliff were floating a barrel of gasoline down the river to the dredge when Jim became aware that Cliff had disappeared. Everyone assumed he had gone for a walk, but at the end of the day he failed to appear and they called the local sheriff. They searched all the places in the river where "things got sucked under," yet no body was discovered. Cliff's marriage itself had apparently been on the rocks, and Jim concluded that he simply wanted to disappear. His last known moments, captured on film, show an intense young man dredging for gold and helping to drag the motorboat over a hill.

Thirty years later at my request, Jim dragged the film reels out of the attic. Sifting through the aimless hours of rocks and river and dredging, I found it hard to imagine what kind of cinematic masterpiece he had in mind. What was the story? Where were the plot points? Who was the audience? How would it be distributed? The questions floated away in my mind like the pack down the river, and, as in the film, nobody involved seems the least concerned.

Chapter Six

CIRCLON SCIENCE

FINANCIALLY SPEAKING, GOLD mining was probably never a viable profession for a man with a family to raise, and in 1970 the couple had settled on the island of Catalina, where Jim began pursuing a parallel career as an abalone diver. Catalina also was a place largely untouched by the forces of industrialization, a haven of the wildness just twenty-six miles from Los Angeles. On the journey over by ferry, you can sometimes see dolphins zooming along by the side of the boat, and these gentle creatures would turn out to be a link to the new theory of matter Jim was in the process of developing. During the next few years on the island, he would crack "the secret" of atomic structure he had begun to think about on those long dark nights at the Blue Hole, and in 1973, '74, and '75, he would publish a series of books laying down the foundation for a complete alternative to the "standard model" of physics.

In the years since the Blue Hole expedition, the world itself had gone through seismic changes. The Vietnam War was at its peak, and American college campuses were exploding in protest. Taking their cue from the civil rights movement, student activists had mobilized by the millions, creating some of the world's largest displays of public opposition to military power. The atrocity of napalm had been witnessed on the nightly news, along with body bags of dead young soldiers, and "draft dodgers" were heroes to a generation who now looked on the war as a violation of international law and a threat to their personal freedom. Women also were taking to the streets, demanding access to education, reproductive choice, and the right to full participation in the workforce. In popular culture, the mood was shifting to a darker tone: In 1973, David Bowie retired his "Ziggy Stardust" persona, and Pink Floyd released *Dark Side of the Moon*; Black Sabbath, Deep Purple, Led Zeppelin, and the Grateful Dead were performing megaconcerts across the United States, while in Sydney two Scottish brothers formed a band called AC/DC. On Catalina, Jim Carter had finally struck gold. With abalone—white and rubbery and hard to get at—Jim at last found financial reward in physically demanding labor.

When Europeans had arrived in North America, banks of abalone stretched for much of the length of the California coast, but by 1970 culling had severely reduced their range. Catalina and its environs remained one of the last great stores of the meat, and restaurants from New York to Tokyo couldn't get it fast enough. Abalone diving was a lifestyle tailor-made to Jim's "yang" temperament, combining, as it did, the technical challenges of scuba equipment with vigorous labor in a remote and beautiful place. Most important, the abalone business provided Jim with a flexible working environment, giving him the freedom to decide when and on what terms he would work. The pay was so good that after a week on an abalone boat he could have several weeks not working, and he used this time to think about the foundations underlying the periodic table of elements. Einstein had once said that the perfect job for a theoretical physicist would be lighthouse keeping, and I like to think he might have viewed abalone diving as a reasonable second choice.

In 1999, I returned with Jim and Linda to the island to see for myself the birthplace of the theory of Circlon Synchronicity. On clear days you can see Catalina from the mainland, yet many Los Angelinos have never visited and even today it remains for most Californians a rather remote dream. When approached by ferry, the island rises out of the water fringed by white-sand beaches and to this day, much of the land remains a private wilderness owned and managed by the Wrigley chewing gum family. With the disappearance of the commercial abalone industry, tourism is now the primary source of revenue and in the sleepy little town of Avalon, bed-and-breakfasts abound. Nestling around the harbor where the ferry comes in, Avalon has strict building codes designed to retain the town's folksy, old-world charm, what the local Chamber of Commerce likes to refer to as Catalina's "timelessness." Around the shoreline a boardwalk planted with palm trees curves around the bay toward a fantastical circular building whose white-tiered archways rise above a pier like a wedding cake. This used to be a casino, and in the 1920s movie stars would come out here to gamble. In 1999 the casino was still the town's tallest building, though it had now been turned into a cinema. Parked in the harbor lay a fleet of yachts, and it was easy to see why anyone with a theory to develop would choose to make this home.

When Jim and Linda had lived here, there were fewer fancy yachts and a lot more working boats; Avalon was home then to an active fishing community. On one of Jim's books from this period, the back cover sports a photo of its author taken from atop the hill overlooking Avalon Bay, and I asked Jim to show me the spot. On a sparkling clear late summer day, he and Linda and my husband and I spent an afternoon winding up the road that leads through town to the crest of the hill, with Jim reminiscing along the way about the early days of his science. As we ascended, he kept up a running commentary on the houses we passed, for the most part quaint little bungalows and cottages, occasionally interspersed with an architectural dream. Few of the houses had changed much since the seventies, Jim said, though real estate prices were skyrocketing and in the hills surrounding the harbor developers were beginning to build pricey condominiums. Jim himself was showing signs of the passing years—his hair had receded and his stomach was no longer the muscular board so impressively on view in the Feather River film—still, in the face that looked out across the harbor when we reached the top of the rise, I could easily detect the young man from the author photo of 1973. As he recounted his time on the island, Jim walked us back to his youth:

> It's hard to remember a time when I wasn't working on physics. But when I was in Catalina, I had a lot more time to devote to it. Basically, that's why I wanted the abalone diving lifestyle—I could go out to the islands for a week or two and make enough money that I didn't have to work for a month. I could work on my theory during bad weather and

> I had all the free time I needed to do what I needed to do. When I was diving I had a lot of time out on the boat at night just to read up on things and to work out my theory, often while I was lying in the bunk at Seal Cove on San Clemente Island.
>
> I did a lot of my best work in Seal Cove when it was too rough to dive. Sometimes the weather would kick up and it would get so bad that you couldn't get out of Seal Cove, so we'd have to sit there for two or three days in the boat. You couldn't dive, couldn't do anything, but I could do physics.

That early photo is the only time Jim has placed an image of himself in his books, and what a wonderfully evocative portrait it is of a physicist-in-the-making. In the grainy black-and-white shot, the scenery has the unhurried languor of seaside towns everywhere, and nothing could seem further from the high-octane world of professional particle science. Unshaved and sporting a mustache, Jim leans toward the camera, holding a cigarillo; his sun-blond hair whips in the wind, and a lock falls rakishly across one eye. The photo has been cropped rather strangely so that the hand holding the cigarillo appears to be breaking the picture frame as in one of those oddly off-kilter perspectival illusions from the fifteenth century. As in Renaissance painting, the effect here is to make the subject feel palpably present; Jim seems almost to be *there* in front of you. The illusion is enhanced by the surrounding graphics, which show a sort of sunburst exploding around the photographic frame. The vanishing point of the sunburst coincides with Jim's head, serving to suggest that this is the point from which the world inside

Figure 6. Back cover of Jim's book Gravitation Does Not Exist. *Photo taken in 1973 at the top of the hill overlooking Avalon Bay on Catalina.* (Jim Carter)

the image—and by implication, the world inside the book—originates. All authors are creators of worlds, and theoretical physicists too are in the business of constructing worlds.

On Catalina, Jim was exploring the question he first encountered through Steve Loftness: What is it that gives atoms structure and makes them fall so naturally into the rows and columns of the periodic table? At the Blue Hole, he had become convinced that atoms were composed of some kind of building blocks that fit together like "subatomic Legos": What are these building blocks? In 1974 Jim published the first part of his answer in a book he titled *The Cosmic Ring*, which offered, as it boldly asserted on its opening page, "a description of reality that is quite different from that generally accepted by contemporary physicists. In fact," the text continued, "some of the ideas presented here dispute the very assumptions upon which all of modern physics is based."

Even before one opened the book, it telegraphed its unorthodoxy. On the bright yellow front cover a collage of coiled springs was arranged into a bull's-eye pattern, like a beacon radiating energy. Was "the cosmic ring" perhaps a call to outer space? Like a lot of graphics from the early seventies, the image was raw and handmade, a crude profusion of vitality. By 1974, the paste-up aesthetic of the protest era was beginning to give birth to the in-your-face visuals of punk rock. In New York the Ramones would give their first performances, and the following year in the UK the Sex Pistols would unleash their brand of anarchy as an inspiration to unschooled genius everywhere. Around the world, guitar players, fashion designers, graphic artists, and filmmakers with no credentials would launch them-

Figure 7. Front cover of Jim's 1974 book The Cosmic Ring. (Jim Carter)

selves like a fleet of AWOL missiles. Far from the epicenters of these proletarian triumphs, Jim Carter was channeling the ethos of the time through the lens of theoretical physics, mounting a one-man assault on the foundations of science with the hubris of a King's Road punk. At 144 pages, stapled together with hand-pasted diagrams and hand-lettered equations, *The Cosmic Ring* was the scientific equivalent of a rock world 'zine. Of all Jim's works, this is my out-and-out favorite.

In his introduction to the book, Jim explained the project: Over the past century, he declared, physics had been invaded by "the virus" of quantum theory, a basket of abstractions that had mired the science in a "mathematical abyss." As Jim saw it, we desperately needed new tools to light our way. The revolution he proposed in *The Cosmic Ring* was a mechanical description of matter and electromagnetic energy, which reimagined our conception of both along more tangible lines. The claims were huge, the book was slim, as Jim was the first to admit. "Perhaps the most conspicuous feature of this book is its brevity, for the subject that it describes could fill several volumes," he wrote. A fully fleshed-out theory of matter would certainly require something more substantial. "I am not, at this time, in a financial position to publish as large a book as this subject deserves," Jim avowed. "Also, some of the ideas in this book took a long time to think up, and since theoretical physics is a hobby with me, I simply do not have as much time as I might like to devote to such pursuits." Our indulgence is begged. Further installments are promised in the future. The first edition of *The Cosmic Ring* must be seen as the introduction, or primer, to a fully developed science.

As one flicks through *The Cosmic Ring*, the first impression is

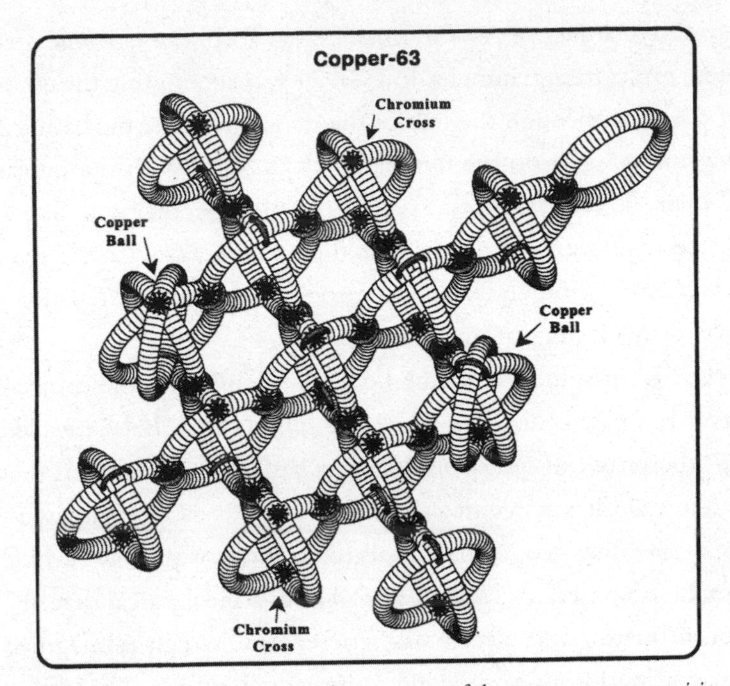

Figure 8. Diagram showing the nuclear structure of the copper atom as envisioned by Jim's theory of Circlon Synchronicity. Here, circlon-shaped particles link together in a subatomic mesh that mirrors the pattern of the Periodic Table of Elements. (Jim Carter)

of graphic exuberance. Diagrams and collages of coiled springs ooze out of the pages. Most of the springs are looped into circles, and many are captioned with names, as if Carter has identified an almost biological taxonomy. Among the various spring "species" are the cosmilons, the gammalons, the violons, the worlons, and the solons. Each spring-type is associated with a particular part of the electromagnetic spectrum and is illustrated by a photo of an intrinsically related object. Thus the ralons, corresponding to what most physicists call "radio waves," are represented by a picture of a radio telescope dish. The excirclons, corresponding to X-rays, are accompanied by an X-ray of a human hand. The redlons, or "infrared radiation," are illustrated by the heat-generating organ of the human brain.

Once you start to look at all the diagrams and charts and tables that Carter has produced, the labor involved is prodigious. It is as if a theory of the universe has been poured forth whole, not in piecemeal parts, as most of the history of science has been done, but in one great creatively joyous swoop. Minutely detailed and obsessively organized, every page is crowded with facts and figures, cross-referencing one another in a system of dazzling complexity. As with the best outsider art, each image seems to rest on the authority of a complete worldview. Crowding at once the space of the page and the mind of the reader, this totalizing aesthetic would define Jim's work for the next thirty years, during which time he would transition from cut-and-paste collaging to software drawing packages, and finally, in the late 1990s, to the motive power of computer animation. But in these early works there is a quality that showcases the raw spirit of the enterprise that in his more mature books would increasingly be muted by efforts to simulate the tone of academic textbooks. There is nothing of the textbook about *The Cosmic Ring*—Jim's creative juices here are bursting at the seams.

"All great truths begin as blasphemies," wrote George Bernard Shaw. As far as mainstream physics is concerned, *The Cosmic Ring* is premised on a blasphemy. For the past century, a major goal of theoretical physics has been to explain physical phenomena in terms of *fields*—physicists speak about magnetic fields, electric fields, and gravitational fields, and at the subatomic level they speak of quantum fields. Many efforts to formulate a "Theory of Everything" are based on the idea that there must be one universal field ultimately responsible for everything. Field theory is not just part of modern physics, it is the foundation of how physicists understand reality. Yet fields are what Jim proposed to jettison, and he understood that by challenging this idea, he was attacking physics at its core.

Ironically, fields themselves had originally been seen as a blasphemous concept, antithetical to the very core of what physics supposedly stood for. The origin of the idea can be found in an experiment now performed by schoolchildren the world over, the simple act of sprinkling iron filings on a sheet of paper near a magnet. As every child now knows, the filings are drawn into a pattern of lines that arc between the magnet's poles, making visible a force that pervades the surrounding space. Michael Faraday, the first person to perform this experiment, was an English physicist who in the 1820s and 1830s pioneered much of what became the foundations of modern electromagnetic theory. Faraday's demonstration that the space around a magnet is pregnant with power struck at the foundation of nineteenth-century physicists' worldview, for nothing in the lexicon of science then could account for this invisible network of lines. Somehow the magnet was exuding an influence that the iron filings could "feel." We take this pattern for granted today as a staple of grade-school science, but stop for a moment to consider what a miraculous phenomenon it is. To scientists of the 1830s, it seemed little short of magic.

Like Jim Carter, Michael Faraday was a practical man without the advantage of a university education. The son of a black-

Figure 9. Michael Faraday (AIP Emilio Segrè Visual Archives, E. Scott Barr Collection)

smith and a child of poverty, he raised himself up by his bootstraps. In his early teens, Faraday had been apprenticed to a bookbinder and seemed destined to spend his life binding other people's ideas into handsome volumes, but when the great chemist Humphry Davy gave a series of public lectures, Faraday bound his lecture notes and sent them to Davy in a gesture of admiration. In return, Davy offered the young enthusiast a job as a bottle washer in his laboratory at the Royal Institution. To Davy's chagrin, the bottle boy would eclipse him, and in many physicists' eyes Faraday would become the greatest experimental physicist of all time.

Because he lacked an academic education, a university

appointment was never an option for Faraday, and he spent his entire career at the Royal Institution, where he eventually became resident professor of chemistry. The RI's mission was to educate the public about the rapidly developing areas of science and technology, and part of Faraday's job was to put on public demonstrations of the latest discoveries in physics and chemistry. His beautifully crafted lectures became the toast of London society, drawing white-tied toffs and satin-gowned ladies along with workingmen and -women. Much like the Society for the Diffusion of Useful Knowledge, the Royal Institution had been formed with the aim of democratizing knowledge and no one could be more representative of how important that goal might be than Faraday, a boy from the slums who rose to become the most beloved scientist in Great Britain, the so-called Cinderella of science.[1]

Faraday's research resulted in his discovery of the principles underlying the electric motor and generator and thereby made possible the muscular miracle of the electrical power industry, yet his experiences with magnets also led him to introduce one of the more ethereal concepts in the scientific pantheon. In thinking about the power around a magnet, Faraday began to imagine that the atoms within the bar were acting like a series of miniature fountains, spraying out an invisible force like millions of tiny geysers. Collectively, the effect was to produce what he called a "force field." While we humans cannot see this field directly, its presence can be detected by the lines of iron filings and the needle of a compass. Faraday discovered that when he held a compass near an active electrical wire, the compass needle would turn, indicating a relationship between electric and magnetic phenomena. Thinking about this relationship, he pro-

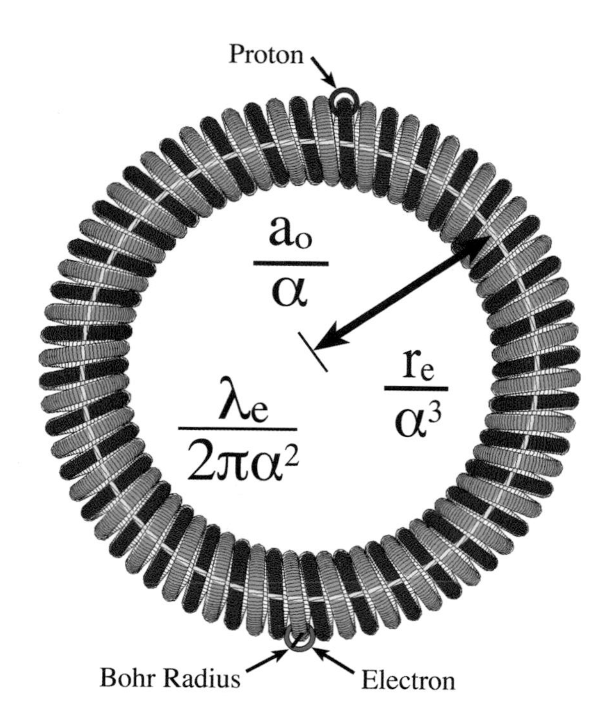

The basic structure of a circlon-shaped subatomic particle in Jim's theory of Circlon Synchronicity. (Jim Carter)

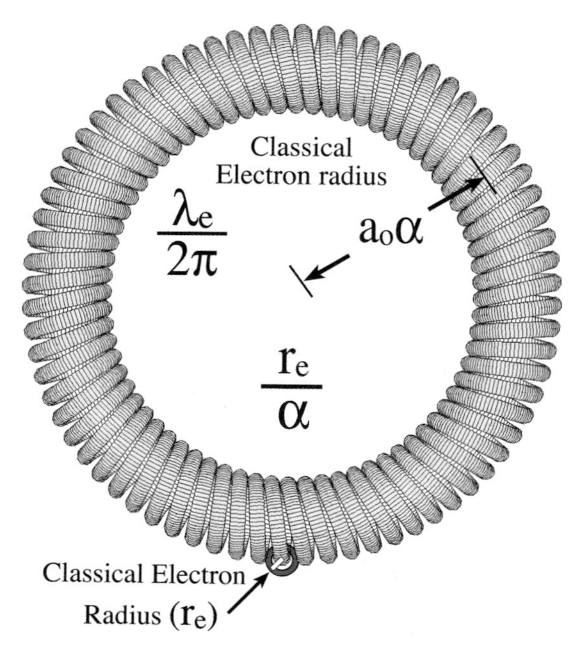

The structure of an electron. (Jim Carter)

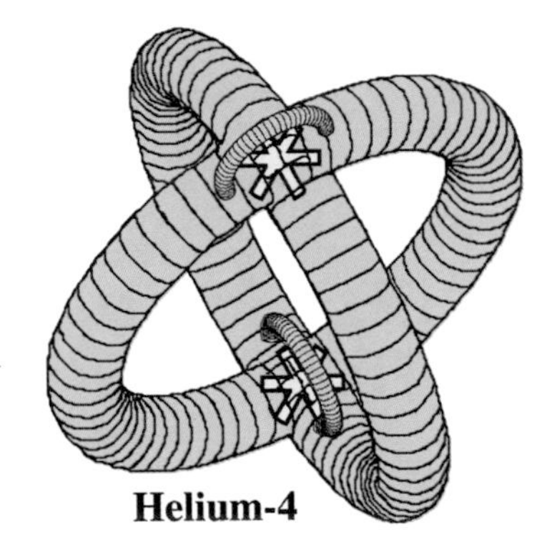

The nucleus of a helium atom. Circlon-shaped particles link together to form a subatomic mesh. (Jim Carter)

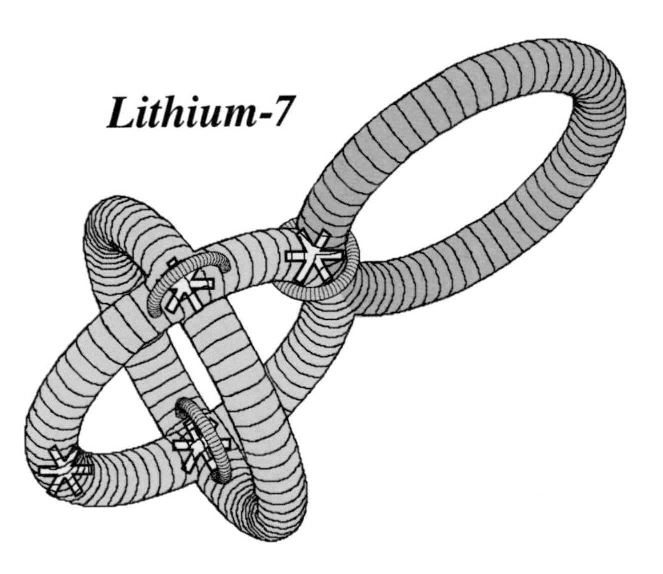

The nucleus of a lithium atom, the element after helium in the periodic table. (Jim Carter)

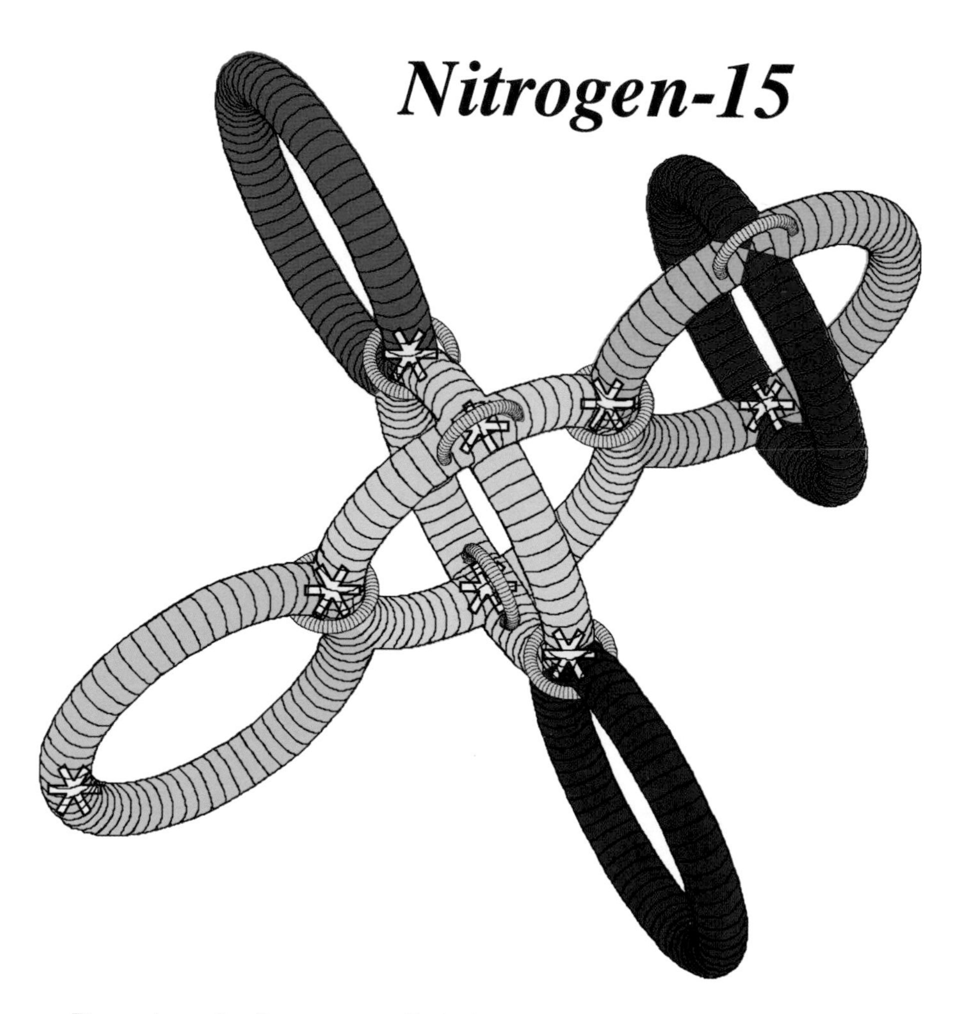

The nucleus of a nitrogen atom. Each element in the periodic table builds up from the addition of more circlon-shaped particles. (Jim Carter)

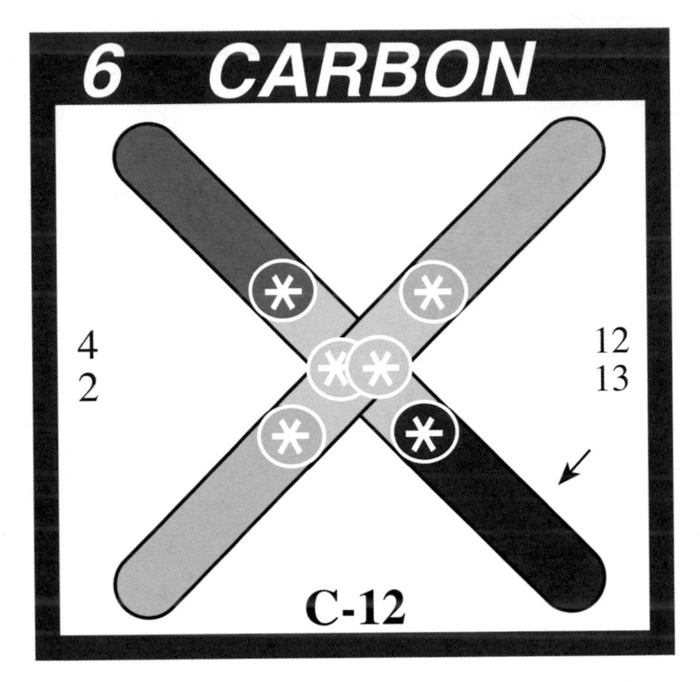

Schematic diagram of the carbon nucleus. (Jim Carter)

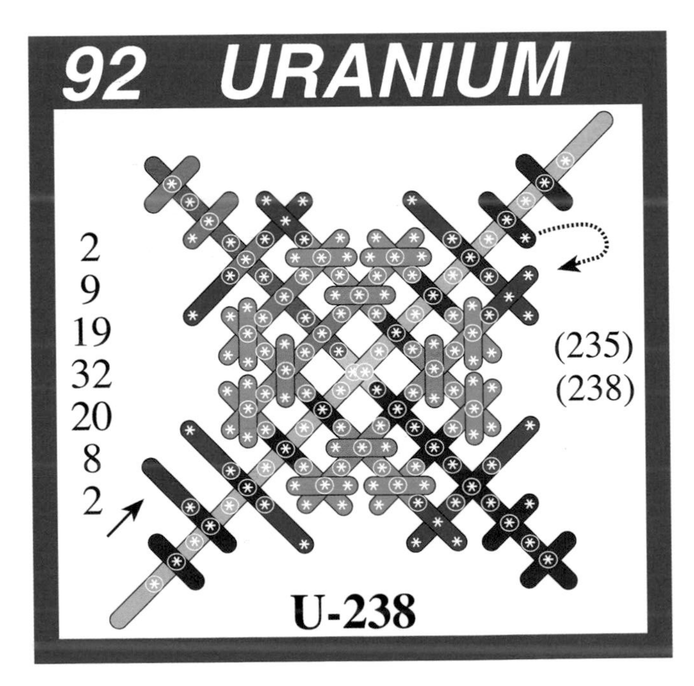

Schematic diagram of the uranium nucleus. (Jim Carter)

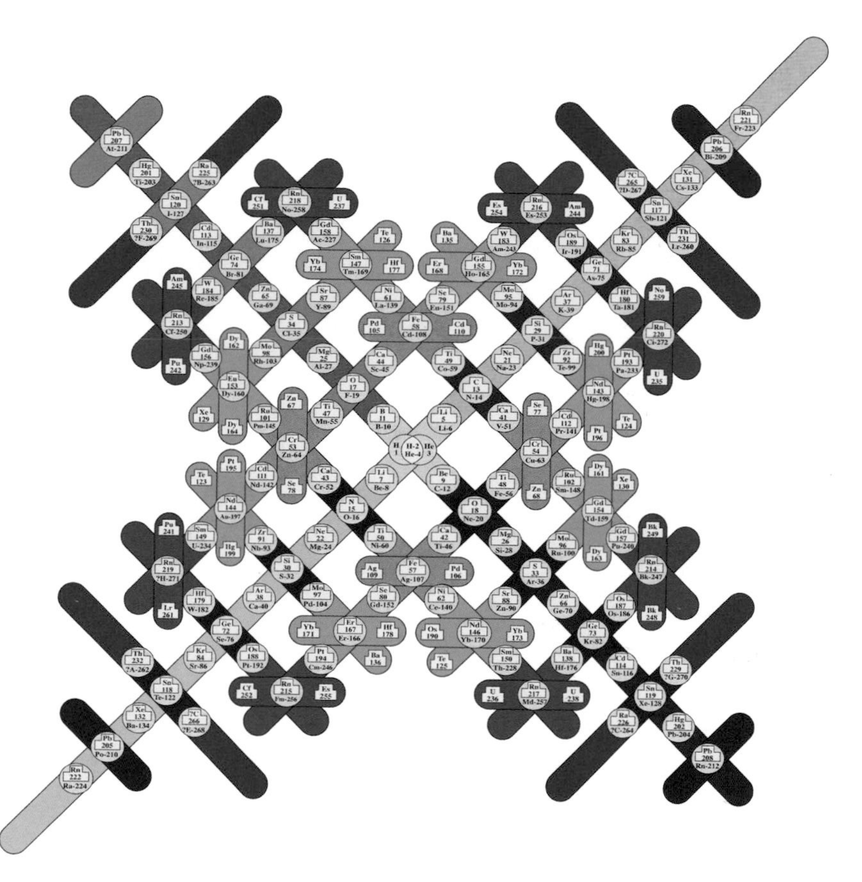

Schematic diagram showing all the elements of the periodic table and their various isotopes. (Jim Carter)

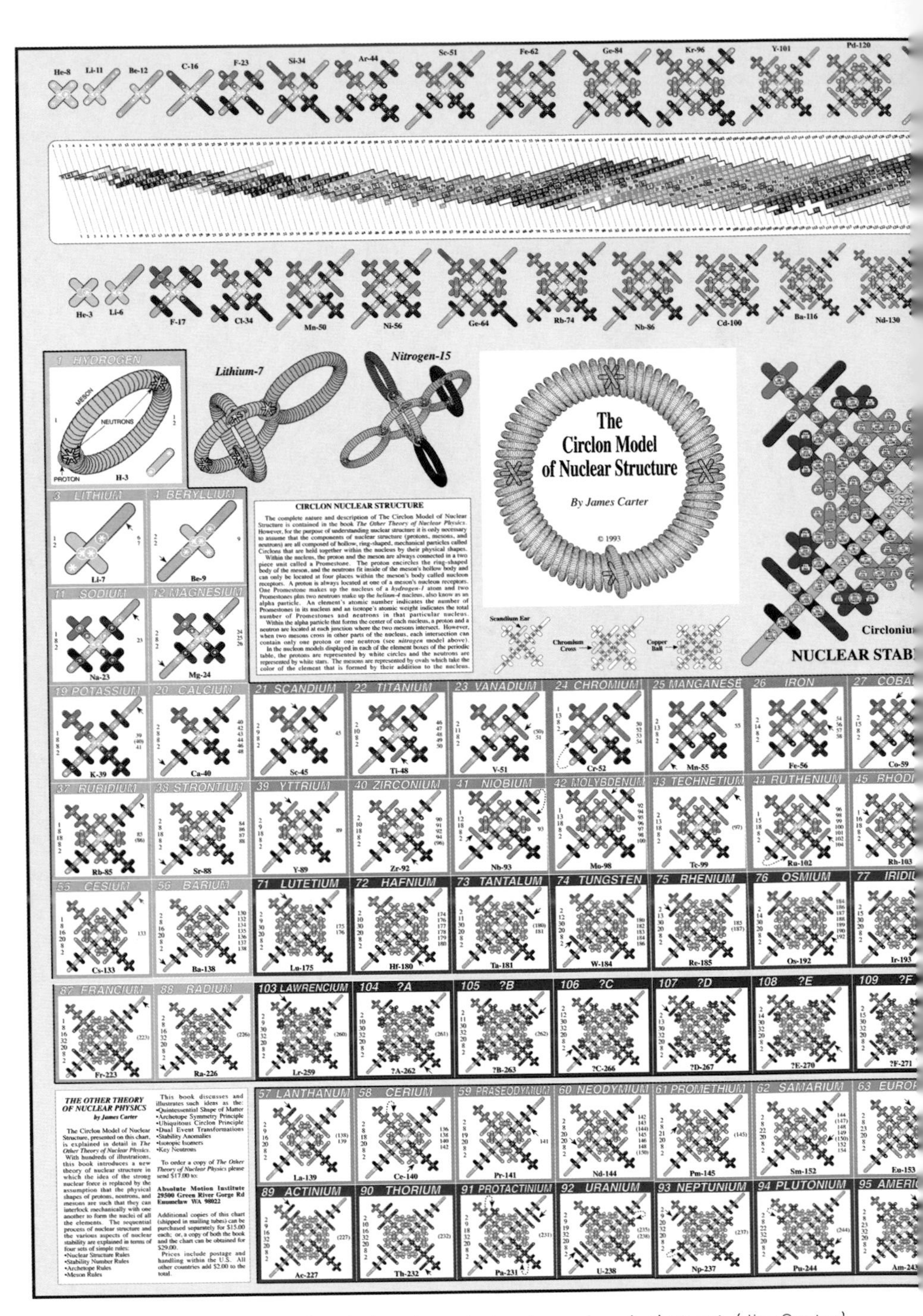

The periodic table, showing the circlon-based structure of each element. (Jim Carter)

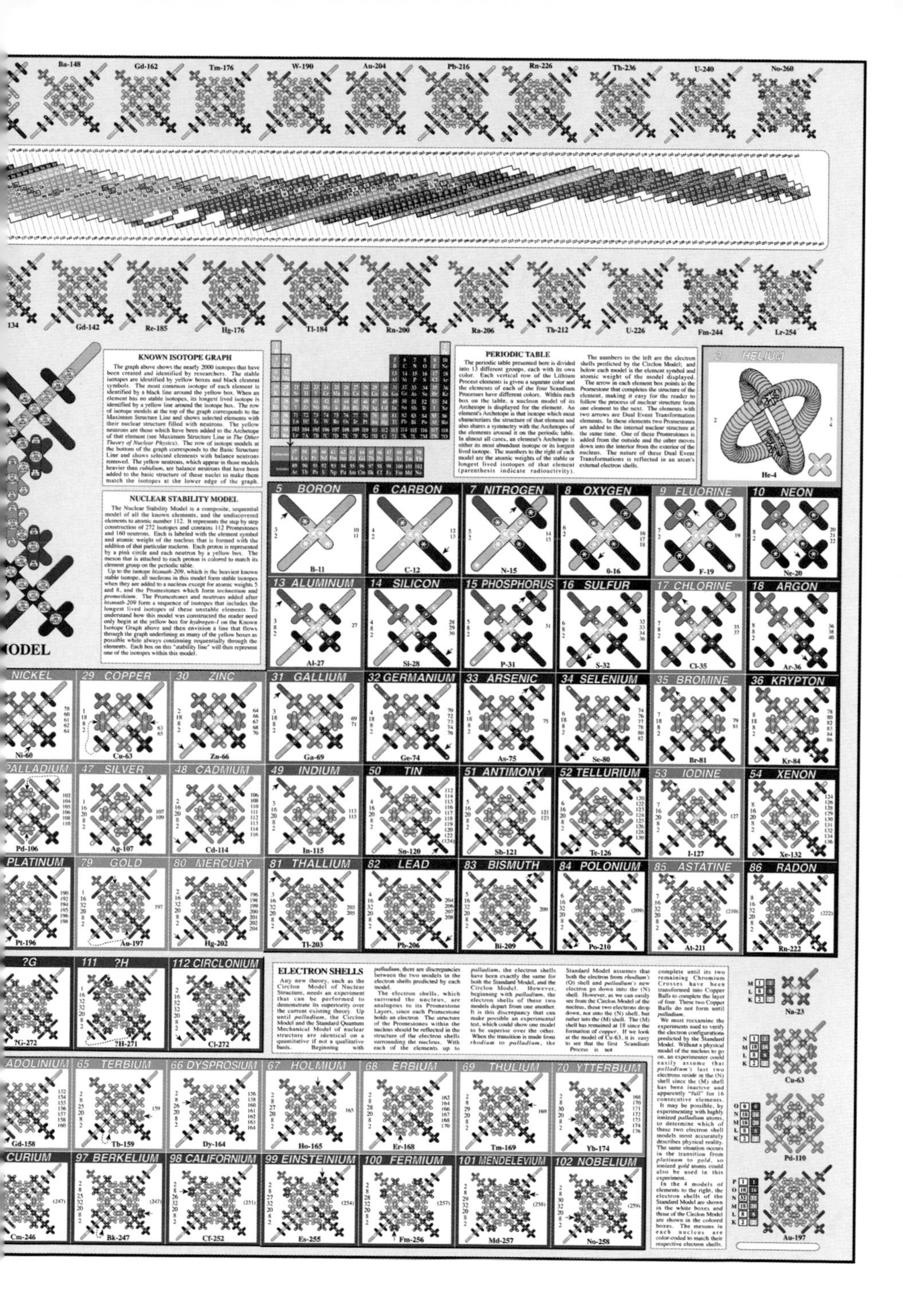

Ba-148 Gd-162 Tm-176 W-190 Au-204 Pb-216 Rn-226 Th-236 U-240 No-260
134 Gd-142 Re-185 Hg-176 Tl-184 Rn-200 Ra-206 Th-212 U-226 Fm-244 Lr-254
KNOWN ISOTOPE GRAPH
The graph above shows the nearly 2000 isotopes that have been created and identified by researchers. The stable isotopes are identified by yellow boxes and black element symbols. The most common isotope of each element is identified by a black line around the yellow box. When an element has no stable isotopes, its longest lived isotope is identified by a yellow line around the isotope box. The row of isotope models at the top of the graph corresponds to the Maximum Structure Line and shows selected elements with their nuclear structure filled with neutrons. The yellow neutrons are those which have been added to the Archetope of that element (see Maximum Structure Line in The Other Theory of Nuclear Physics). The row of isotope models at the bottom of the graph corresponds to the Basic Structure Line and shows selected elements with balance neutrons removed. The yellow neutrons, which appear in those models heavier than rubidium, are balance neutrons that have been added to the basic structure of these nuclei to make them match the isotopes at the lower edge of the graph.
NUCLEAR STABILITY MODEL
The Nuclear Stability Model is a composite, sequential model of all the known elements, and the undiscovered elements to atomic number 112. It represents the step by step construction of 272 isotopes and contains 112 Promestones and 160 neutrons. Each is labeled with the element symbol and atomic weight of the nucleus that is formed with the addition of that particular nucleon. Each proton is represented by a pink circle and each neutron by a yellow box. The meson that is attached to each proton is colored to match its element group on the periodic table.
Up to the isotope bismuth-209, which is the heaviest known stable isotope, all nucleons in this model form stable isotopes when they are added to a nucleus except for atomic weights 5 and 8, and the Promestones which form technetium and promethium. The Promestones and neutrons added after bismuth-209 form a sequence of isotopes that includes the longest lived isotopes of these unstable elements. To understand how this model was constructed the reader need only begin at the yellow box for hydrogen-1 on the Known Isotope Graph above and then envision a line that flows through the graph underlining as many of the yellow boxes as possible while always continuing sequentially through the elements. Each box on this "stability line" will then represent one of the isotopes within this model.
ODEL
PERIODIC TABLE
The periodic table presented here is divided into 13 different groups, each with its own color. Each vertical row of the Lithium Process elements is given a separate color and the elements of each of the four Scandium Processes have different colors. Within each box on the table, a nucleon model of its Archetope is displayed for the element. An element's Archetope is that isotope which most characterizes the structure of that element and also shares a symmetry with the Archetopes of the elements around it on the periodic table. In almost all cases, an element's Archetope is either its most abundant isotope or its longest lived isotope. The numbers to the right of each model are the atomic weights of the stable or longest lived isotopes of that element (parenthesis indicate radioactivity).
The numbers to the left are the electron shells predicted by the Circlon Model; and below each model is the element symbol and atomic weight of the model displayed.
The arrow in each element box points to the Promestone that completes the structure of the element, making it easy for the reader to follow the process of nuclear structure from one element to the next. The elements with two arrows are Dual Event Transformation elements. In these elements two Promestones are added to the internal nuclear structure at the same time. One of these Promestones is added from the outside and the other moves down into the interior from the exterior of the nucleus. The nature of these Dual Event Transformations is reflected in an atom's external electron shells.
2 HELIUM
He-4
5 BORON B-11
6 CARBON C-12
7 NITROGEN N-15
8 OXYGEN O-16
9 FLUORINE F-19
10 NEON Ne-20
13 ALUMINUM Al-27
14 SILICON Si-28
15 PHOSPHORUS P-31
16 SULFUR S-32
17 CHLORINE Cl-35
18 ARGON Ar-36
NICKEL Ni-60
29 COPPER Cu-63
30 ZINC Zn-66
31 GALLIUM Ga-69
32 GERMANIUM Ge-74
33 ARSENIC As-75
34 SELENIUM Se-80
35 BROMINE Br-81
36 KRYPTON Kr-84
PALLADIUM Pd-106
47 SILVER Ag-107
48 CADMIUM Cd-114
49 INDIUM In-115
50 TIN Sn-120
51 ANTIMONY Sb-121
52 TELLURIUM Te-126
53 IODINE I-127
54 XENON Xe-132
PLATINUM Pt-196
79 GOLD Au-197
80 MERCURY Hg-202
81 THALLIUM Tl-203
82 LEAD Pb-206
83 BISMUTH Bi-209
84 POLONIUM Po-210
85 ASTATINE At-211
86 RADON Rn-222
?G ?G-272
111 ?H ?H-271
112 CIRCLONIUM Cl-272
ELECTRON SHELLS
Any new theory, such as the Circlon Model of Nuclear Structure, needs an experiment that can be performed to demonstrate its superiority over the current existing theory. Up until palladium, the Circlon Model and the Standard Quantum Mechanical Model of nuclear structure are identical on a quantitative if not a qualitative basis. Beginning with palladium, there are discrepancies between the two models in the electron shells predicted by each model.
The electron shells, which surround the nucleus, are analogous to its Promestone Layers, since each Promestone holds an electron. The structure of the Promestones within the nucleus should be reflected in the structure of the electron shells surrounding the nucleus. With each of the elements up to palladium, the electron shells have been exactly the same for both the Standard Model, and the Circlon Model. However, beginning with palladium, the electron shells of these two models depart from one another. It is this discrepancy that can make possible an experimental test, which could show one model to be superior over the other.
When the transition is made from rhodium to palladium, the Standard Model assumes that both the electron from rhodium's (O) shell and palladium's new electron go down into the (N) shell. However, as we can easily see from the Circlon Model of the nucleus, these two electrons drop down, not into the (N) shell, but rather into the (M) shell. The (M) shell has remained at 18 since the formation of copper. If we look at the model of Cu-63, it is easy to see that the first Scandium Process is not complete until its two remaining Chromium Crosses have been transformed into Copper Balls to complete the layer of four. These two Copper Balls do not form until palladium.
We must reexamine the experiments used to verify the electron configurations predicted by the Standard Model. Without a physical model of the nucleus to go on, an experimenter could easily assume that palladium's last two electrons reside in the (N) shell since the (M) shell has been inactive and apparently "full" for 16 consecutive elements.
It may be possible, by experimenting with highly ionized palladium atoms, to determine which of these two electron shell models most accurately describes physical reality. The same situation occurs in the transition from platinum to gold, so ionized gold atoms could also be used in this experiment.
In the 4 models of elements to the right, the electron shells of the Standard Model are shown in the white boxes and those of the Circlon Model are shown in the colored boxes. The mesons in each nucleus are color-coded to match their respective electron shells.
Na-23
Cu-63
Pd-110
Au-197
ADOLINIUM Gd-158
65 TERBIUM Tb-159
66 DYSPROSIUM Dy-164
67 HOLMIUM Ho-165
68 ERBIUM Er-168
69 THULIUM Tm-169
70 YTTERBIUM Yb-174
CURIUM Cm-246
97 BERKELIUM Bk-247
98 CALIFORNIUM Cf-252
99 EINSTEINIUM Es-255
100 FERMIUM Fm-256
101 MENDELEVIUM Md-257
102 NOBELIUM No-258

The Green River Gorge with the Carters' house in the background (top right of photo). (Linda Carter)

Peter Guthrie Tait's smoke ring generator—a used packing crate covered with a damp towel. (From *Lectures on Some Recent Advances in Physical Science*, 1876, by P. G. Tait)

Jim's smoke ring generator—a garbage can covered with rubber sheeting. (Linda Carter)

Smoke rings sail across the Carters' front yard. (Linda Carter)

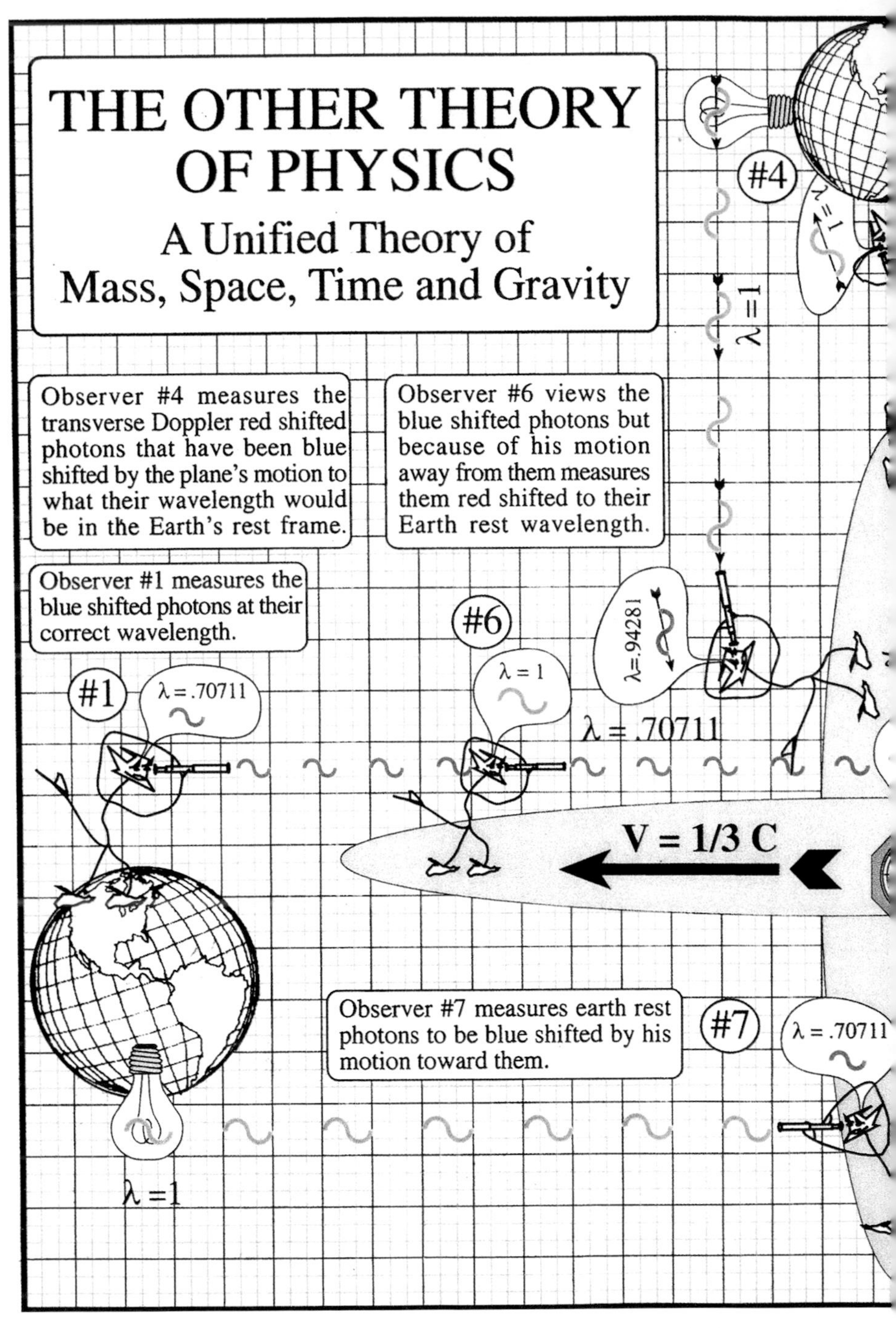

Diagram explaining the astronomical "redshift" of distant stars. Jim's theory proposes an alternative to special relativity. (Jim Carter)

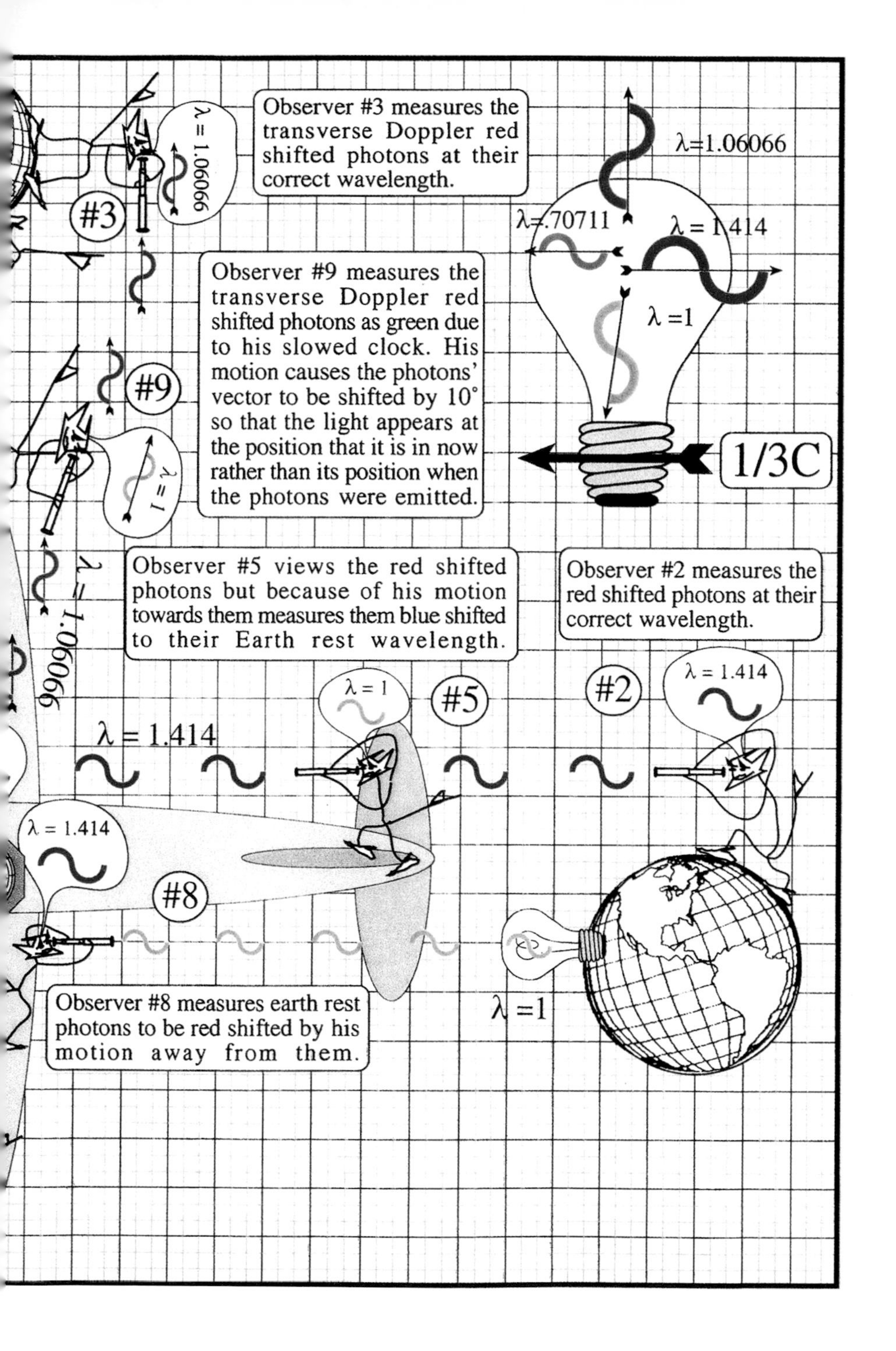
Observer #3 measures the transverse Doppler red shifted photons at their correct wavelength.
λ = 1.06066
#3
λ=1.06066
λ=.70711
λ = 1.414
Observer #9 measures the transverse Doppler red shifted photons as green due to his slowed clock. His motion causes the photons' vector to be shifted by 10° so that the light appears at the position that it is in now rather than its position when the photons were emitted.
λ =1
#9
λ = 1
1/3C
Observer #5 views the red shifted photons but because of his motion towards them measures them blue shifted to their Earth rest wavelength.
Observer #2 measures the red shifted photons at their correct wavelength.
λ = 1.06066
λ = 1
#5
#2
λ = 1.414
λ = 1.414
λ = 1.414
#8
Observer #8 measures earth rest photons to be red shifted by his motion away from them.
λ =1

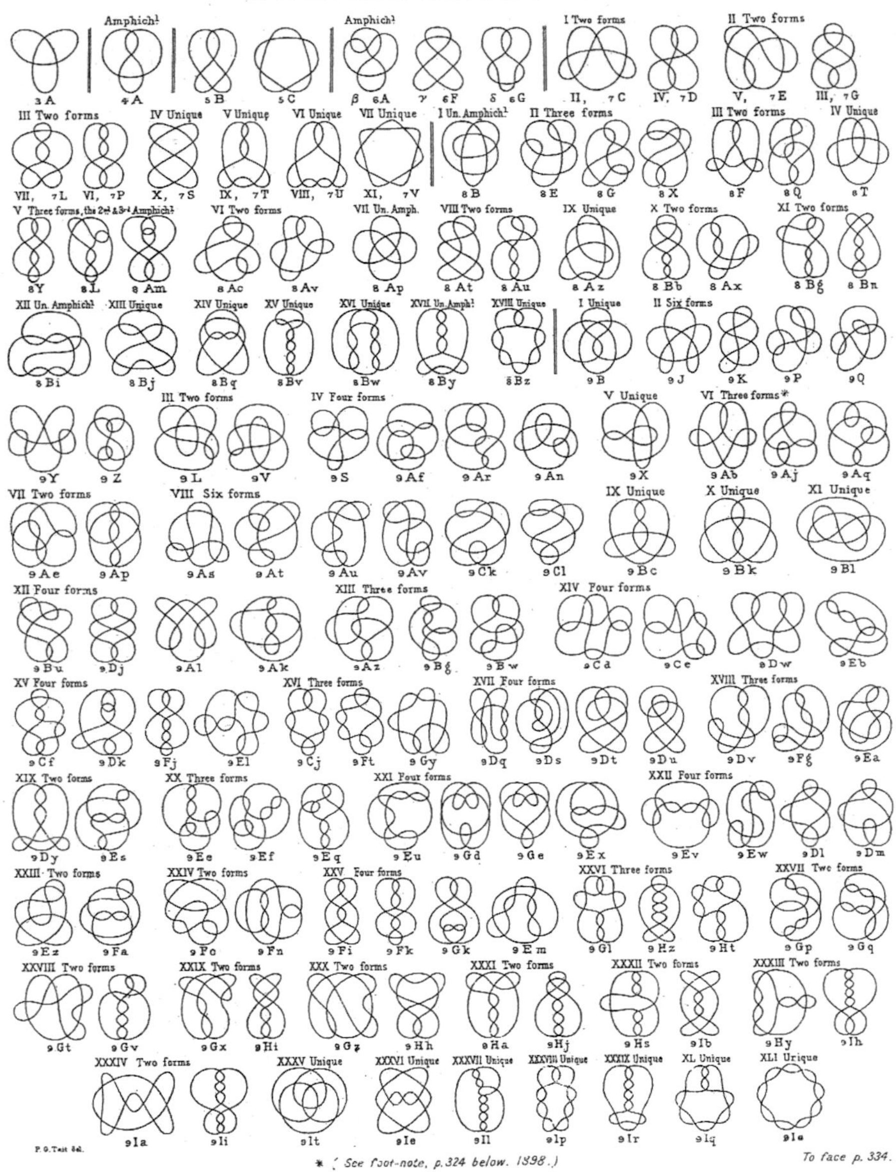

"Knot table" cataloging all the distinct knots up to seven crossings of a string. (From *The Unseen Universe; or, Physical Speculations on a Future State*, 1880, by P. G. Tait)

Jim with a fresh catch of abalone on Catalina Island. (Carter family archives)

An early version of Jim's lift bags. (Carter family archives)

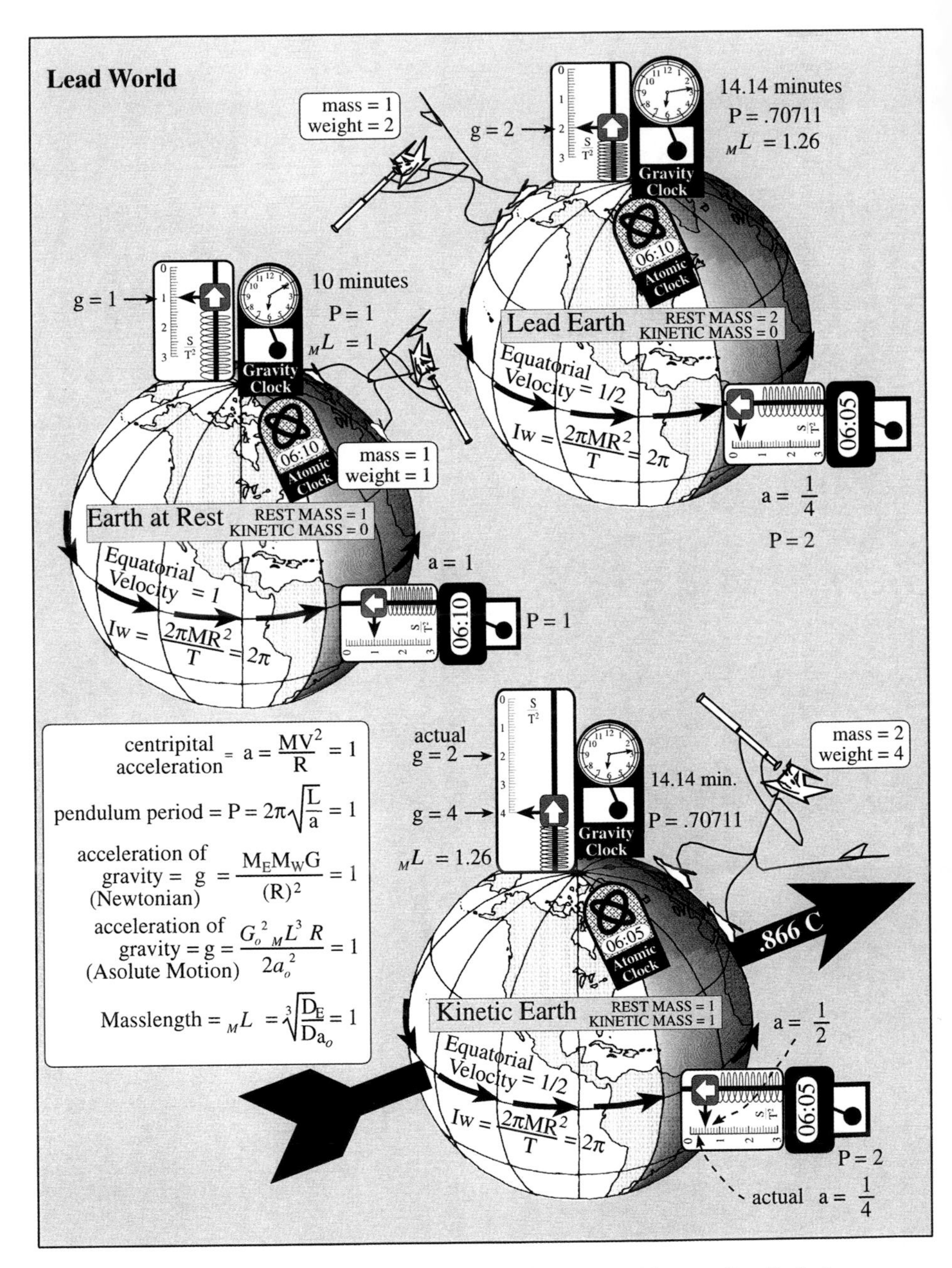

Diagram describing apparent variations in time caused by gravity. Jim's theory proposes an alternative to general relativity. (Jim Carter)

Jim's "Creation Spiral," showing the number of particles in the universe as they split off from two original circlon-shaped particles. (Jim Carter)

Jim's original plywood models of the circlon-based structures of atomic nuclei. (Linda Carter)

Jim Carter, May 2011. (Linda Carter)

posed that the electric force also operates via an electric field. And Faraday went further, suggesting that the force of gravity was due to a gravitational field. He even proposed that all three fields were different manifestations of one ultimate field, an extraordinary insight for the time. But Faraday's fields challenged the mechanistic worldview that had reigned in science since the late seventeenth century, and in the assertively industrial age of the mid–nineteenth century, when men of science were in the business of helping to improve steam engines, the idea did not go down well. To many professional physicists fields smacked of magic in much the same way that Newton's notion of a gravitational *force* had done two centuries before. The fact that Faraday didn't have a degree did not help his case, and it has been said by one of his biographers that he died of a broken heart because his beloved idea was not taken seriously by his scientific peers.

In the 1860s, however, fields acquired a champion in the form of James Clerk Maxwell, the Newton of the nineteenth century. Maxwell framed Faraday's concept in mathematical terms with a set of equations that described in quantitative terms the precise interaction between magnetic and electric phenomena. Maxwell's equations remain perhaps the most useful ever written down; the telecommunications industry was spawned by his prediction of what came to be called "radio" waves, and the electricity coming into our homes is borne on the application of these equations. The success of Maxwell's equations convinced most physicists to accept the field concept, and in the late nineteenth century field theory ushered in an intellectual revolution that transformed the way physics was done. General relativity, quantum theory, and string theory are all triumphs of field theory. By the time Jim was writing *The Cosmic Ring*, Faraday's heresy had become orthodoxy and physicists had developed field theories to explain *all* of nature's fundamental forces. Moreover, they had come to regard matter itself as a manifestation of quantum fields. Every kind of subatomic particle—the electron, the quark, and so on—is now said to be associated with its own quantum field. In this way of seeing, particles of matter are indeed nothing more than localized peaks of intensity in some ethereal quantum field.

Fields changed the way physicists understood the world at its most basic level by literally dematerializing matter. In many ways, field theory revolutionized physics more profoundly than quantum theory itself. In the sixteenth century, Copernicus had challenged our conceptions of *where* we are; Faraday's fields challenged received wisdom about *what* we are—not conglomerations of hard little balls, but shimmering networks of energy. By the start of the twentieth century, the field-based worldview had replaced the mechanistic paradigm, and its power has been enhanced ever since. Today, physicists describe fields using complex mathematical tools such as Lie algebras and gauge symmetries, and they propose that fields can be twisted into topological knots in multidimensional space.

Jim Carter, lying in his bunk on an abalone boat off the coast of Catalina, found this account of reality hard to accept. His intellect rebelled against the abstraction of fields; their mathematical abstruseness offended the concrete turn of his mind. Ironically, the one major figure in the history of physics whose life story in some respects paralleled his own had been the source of an idea he could not stomach. Jim was convinced, just as the founders of the scientific revolution had held, that nature was a mechanical system, and his intuition told him that a sensible man with a sensible mind could find a sense-able description of the world. He wanted to take physics back to the pre-field paradigm and ground our understanding in concepts he could *see* and *feel*. He did not imagine it would be a trivial task, yet given that the human brain was also a product of nature's laws neither could it be an insurmountable one.

In his bunk at Seal Cove, Jim searched for a new way to understand the world. Quantum theory had given physicists a description of the atom that explained the pattern of elements in the periodic table, so in order to be taken seriously any alternative theory would have to account for this pattern. How could he explain atoms with a non-field theory? Originally derived from the ancient Greek word *atomos*, meaning indivisible, "atoms" were traditionally viewed as the most basic possible particles; they were the things that could not be further subdivided. Today the term is used to describe things that *can* be subdivided, but that is only because scientists of the nineteenth century hadn't yet realized that the things *they* called atoms were not the smallest possible entities. That status now belongs to subatomic particles such as electrons and quarks, which constitute the basic matter units of physics today.

Whatever we call the smallest things, the same questions arise: What do these units look like? What are they made of? How do they interact with one another? The Greeks could not answer such questions, which led some ancient philosophers to reject the whole idea. Most important, Aristotle didn't believe in atoms and held instead that matter was an infinitely divisible *continuum*. Atoms were one of the ancient ideas that came back into fashion during the seventeenth century, and in the eighteenth century physicists applied themselves to thinking through how they might actually operate.

Logic dictates that if atoms exist, they must be able to combine in a great variety of ways, for how else could we account for the dazzling spectrum of stuff that nature so manifestly makes—everything from air and water and sea foam to wood and marble and emeralds? At the *elemental* level, there must be some minimal set of substances from which all else is made. One of the major scientific challenges of the nineteenth century was to clarify what constituted the basic set of "elements" that could not be further broken down—the oxygen and hydrogen that make up water, the sodium and chlorine that make up salt, and so on. Countless scientists contributed to this effort, and in 1862 an irritable Russian genius named Dmitri Mendeleev completed the task of finding the pattern that organized them as a whole. In Mendeleev's periodic table, the atomic elements are grouped in rows by order of their increasing atomic weight; thus hydrogen with atomic weight 1 is the first element, helium with atomic weight 4 is the second element, lithium with atomic weight 7 is the third. In Mendeleev's table, the columns are organized according to the elements' similar chemical properties; thus the inert "noble gases" are in one column (helium, neon, argon, krypton, xenon, and radon); the highly reactive halogens are in another (fluorine, chlorine, bromine, iodine, and astatine), many of which are damaging to biological organisms and are often used as disinfectants and bleaches. It is the "periodicity" of the columns that gives the table its name. Since Mendeleev's day a few small amendments have been made, but essentially his scheme is the one that now adorns classrooms the world over.

Yet if Mendeleev clarified the *pattern* of the elements, the *substance* of atoms remained obscure. What actually *is* an "atom"? For most of the nineteenth century, scientists thought the atomic elements *were* the most basic units, because these were the things that remained unchanged during chemical reactions. They imagined that hydrogen, oxygen, carbon, and so forth were little indivisible masses that joined together through various rearrangements. At the end of the nineteenth century, however, physicists discovered hints that atoms were themselves composed of even smaller parts. In 1897, J. J. Thomson discovered the "electron." In 1909, the New Zealand–born physicist Ernest Rutherford fired alpha particles at a thin sheet of gold foil and found that atoms had a hard little core at the center of a mostly empty space. This "nucleus" turned out to be made of "protons" and "neutrons." With the discovery of these subatomic particles the goalpost shifted, yet the basic questions remained: What is the nature of *these* parts? What do *they* look like? What are *they* made of?

The standard answer today is that all these particles of matter are localized peaks of intensity in various quantum fields. Just as a magnetic field will be concentrated around a magnet, so a "particle" can be seen as a localized concentration of a field. At the subatomic level, physicists now tell us, solidity is an illusion. In the final analysis of field theory, there is no *substance* at all, only fields that attract and repel one another. Now, of course, we are inclined to ask: What is the nature of these fields? What do *they* look like? What are *they* made of? To physicists, the answer is that they are "states of space." They do not *look* like anything, for they are more primary than the medium of vision, which is light. They are not *made* of anything; they are conditions in the fabric of reality, about which we can say little except that it *is*. As to how fields *behave*—for that we have the equations of quantum field theory, whose predictions have been verified to more than twenty decimal places. If you want to understand matter, a mainstream physicist would say, you have to understand quantum fields, and if you want to understand these, there is no way forward but to learn the math.

In his bunk bed on the abalone boat at Seal Cove, Jim could not believe nature spoke a language so inaccessible to the common man. He did not accept that a Ph.D. was a necessary requirement for a dialogue with the world around him. He was convinced that atoms were composed from some kind of Lego-like units that would piece together in a necessary way to build up the pattern in the periodic table one unit at a time. Alone on the ocean, he toyed mentally with different forms these units might take, letting his mind wander in the space of possibility as he imagined how various shapes could fit together. An element's order in the periodic table is determined by the number of protons in its nucleus: Thus hydrogen, the first element, has one proton; helium, the second element, has two protons; lithium, the third element, has three protons; and so on. Proton count dictates the primary characteristic by which an atom is defined and is known as its atomic number. Protons are accompanied by neutrons, and the number of neutrons determines the "isotope" of the element.

In the standard theory of physics, protons and neutrons pack together in the little ball of the nucleus, and initially Jim tried to imagine his models as ball-like clusters, too, but he could not find a system that worked to match Mendeleev's pattern.

His breakthrough came on a trip home to Buckley in 1971

during the Fourth of July weekend when he was taking his son Paul to meet his parents. This would prove to be one of the seminal experiences of his scientific life. For kids growing up in Buckley, the Fourth of July had always been a major event; as teenagers Jim and his friends would spend the day setting off fireworks and exploding "handmade bombs." On the drive up from Catalina, Jim now conceived a new idea for the bombs: He would "take a plastic soda bottle, drill a hole in the cap just the right size to insert a cannon fuse, pour in a teaspoon of gunpowder, light the fuse, and run." He thought it would be interesting to find out what would be the optimal amount of gunpowder to produce the loudest bang. Though he "wasn't exactly sure what might happen" nothing prepared him for "the miracle that occurred" when the first bomb exploded. Along with the bang, a perfectly formed smoke ring came rising out of the blast and sailed into the sky. Lying on his back in the grass, watching this ring for what seemed an eternity, Jim couldn't believe what he was witnessing. "How could such a beautiful and complex unit of order and symmetry come out intact from such a violent explosion?" What made the ring so stable and kept it so perfectly formed?

More bombs proved that the first wasn't a coincidence—with the right amount of gunpowder, about half of the explosions produced smoke rings—and that night Jim went to bed with his head full of smoke. As he was falling asleep, he was stirred by "a delicious little dream" and suddenly the answer to the puzzle of nuclear structure crystallized in his mind. Smoke rings from the bomb blast had generated themselves; could it be possible that subatomic particles were also miniature toroidal rings? In his mind, Jim began to imagine little clouds of toroids linking together. What if protons and neutrons and other subatomic particles were all different sizes of this basic "circlon" form, interlinking like a sort of chain-mail mesh?

To test his ideas, Jim spent the next morning in his father's workshop with a drill press and a hole saw cutting two-inch-diameter rings out of pressed hardboard. He sanded the rings and painted them black. They were the raw materials for a revolution. On the drive back to Catalina, his mind was abuzz with how he might fit the rings together, and by the time he was home in his "physics room" in Avalon, he knew what he had to do.

Working in a pitch of inspiration, he took a piece of white card and hung one of the black rings in front of it. This represented the nucleus of the hydrogen atom, with atomic number 1. He photographed the ring, then he glued two rings together at right angles, forming a little ball-shaped cross to represent helium, with atomic number 2. He photographed this and added a third ring to represent lithium, atomic number 3. Rapidly he worked his way through the succeeding elements, adding one ring at a time, in what seemed like an almost necessary sequence of forms. At each step he photographed the resulting model, until finally, at element 101, he ran out of rings. The next day he developed the photos and cut up the images into squares that he pasted on top of the corresponding elements on a periodic table on his wall. Incredibly, each row of Mendeleev's table seemed to represent an "obvious layer" in the structure of his circlon models. In the table's vertical rows, "the outer structure of the models was almost identical."

Jim was on a roll, but he still had some blanks to fill in. Specifically, if his model was correct, it would need to explain not just the basic elemental structure, but also the possible isotopes of each element, for many elements have more than one isotope.

To take an example almost everyone knows, uranium has two isotopes: U-235, which is used in nuclear reactors, and the far more common U-238, which is not fissile directly but is used as the breeding stock to generate plutonium. Both isotopes have 92 protons, but U-235 has 143 neutrons, while U-238 has 146 neutrons. Carbon also has several isotopes, including the most common type, C-12, which is the primary chemical of life, and the much rarer C-14, which archaeologists measure with carbon-dating techniques.

Jim had run out of plywood rings, so he sat down and started drawing diagrams of all the possible isotopes of each element. In one sitting he made drawings of almost three hundred known isotopes. "To my great satisfaction," he wrote, "I found that when I constructed the various isotopes for each element, the one [whose model] was the most symmetrical and balanced inevitably turned out to be that element's most abundant isotope." By the time he finished, Jim could predict which ones were radioactive—he could tell by looking at his circlon models whether any isotope was stable. He was twenty-seven years old and he had "discovered some kind of secret of the ages."

Jim did not know it, but the method he used was reminiscent of the one Mendeleev himself had employed. During the years Mendeleev was working out the table, he kept a series of note cards, inscribed with the properties of each element. For decades he carried these cards around with him, arranging and rearranging them into patterns to try to find the structure that made sense of them all. It turns out Mendeleev was a card player with a penchant for the game of patience, which he would play obsessively on long-distance train trips across the Siberian tundra, moving the cards rapidly into their respective suits. One of his biographers has suggested that his success with the periodic table was due at least in part to his long-honed ability to see patterns in decks of playing cards.

Jim was unaware of this historical parallel, but he recognized that with his ring models his search was over. He assembled his photographs and diagrams into a wall chart to make his own version of the table, showing each element's structure according to the rules of his circlon theory. Copies could be purchased by mail order from the Universal Expansion Press; price $3. The charts sold in sufficient quantity that in 1977 he published a second edition, and in 1993 he updated the chart, replacing the original black-and-white images with color-coded graphics.[2]

In 1975, he wrote up his findings in a book he called *The Circlon Atom*, in which each page was devoted to a single element. "Beryllium is a light, strong and elastic metal . . . ," begins one entry. "Boron is a brownish-black crystalline non-metal . . . ," begins another. "Sulfur is a pale-yellow, odorless brittle solid, which is insoluble in water, but soluble in carbon disulfide," another one tells us. In Oliver Sacks's wonderful book about the history of chemistry, *Uncle Tungsten*, he also would give us an elemental rundown. Packed into Jim's book were all the physical and chemical data he could amass about each element, making this slim volume a miniencyclopedia of atomic information. Stripped down graphically, symbolic, and boldly elemental, the images here resemble circuit diagrams or patterns for complex pieces of lace. It is Jim's most elegant work.

With his ring models, Jim had answered to his own satisfaction the question of how atomic nuclei are arranged; he had solved the architectural problem of how to obtain the periodic table of elements from basic Lego-like parts. Now he faced the existential challenge that all atomic theorists must confront: What are these parts made of, and how do they bind together? Jim had made his rings out of plywood, but his inspiration had been smoke rings, and he speculated that the basic units of matter must also be hollow tubes. One of the advantages of tubes is that larger tubes could contain smaller tubes inside them. He reasoned that each basic particle type—the proton, the neutron, the meson, and electron—was a different-sized toroid.

But what were the toroids themselves made of? Here Jim had another of his critical revelations: What if the toroids were not literally tubes, but actually composed from some kind of microscopic "wire" or "string" wound into tight coils? From a distance they would appear to be tubes, but up close they would resemble springs. To test this idea, he set about making a series of springs by winding baling wire around a broom handle. He welded the springs together at their ends to make loops and joined a group of loops together to construct a model of the helium atom.

The advantages of springs were immediately evident, for unlike plywood rings, springs are flexible. The plywood models were fragile and easily broken, whereas springs could flex and bend and bounce back into shape when stressed. Real atoms have the latter qualities. Springs can also vibrate, as we know atoms do. Jim now planned to make the entire periodic table from springs, just as he had done with his rings. He hoped to use spring models to demonstrate the chemical properties of each element and to show how different elements behave in chemical reactions. He planned to write a book on this subject as a companion to *The*

Cosmic Ring, tentatively titling the project *The Crystal Atom*. It "would be to chemistry what *The Cosmic Ring* was to physics."

Unfortunately, making springs was extremely time-consuming, and the labor proved too much even for him. In the introduction to *The Cosmic Ring* Jim explained the holdup on *The Crystal Atom* project: "I am working on this book, but I don't know when I will be able to publish it, for I am becoming bogged down in the sheer drudgery and complexity of making the wire models," he wrote. "To properly illustrate something like uranium fission would take over 1,000 hand-made wire circlons. Even something rather simple like a water molecule takes a long time to make." He ended with a plea for assistance. In return for a modest donation, he promised to send donors further editions of his books. It was the only time I have known Jim to ask for support. If the foundations of physics needed shoring up, he was not a man to sit around waiting for a grant from the National Science Foundation in order to begin. Like digging for treasure, the Ultimate Theory of Reality was something you had to be willing "to slave" for.

Many years later, Jim described to me his method of scientific theorizing, an activity he likened to his exploits on the Feather River. Although few academic scientists may have chosen his metaphor, it is one I suspect many of them would agree with. "Theoretical physics," he explained, "is a lot like gold mining—you have to dig away a lot of dirt to get to those nuggets of gold."

Not nuggets of gold, but coils of baling wire: That was the result of Jim's years on Catalina. According to his theory, everything in our universe results from the actions and interactions of springy, circlon-shaped particles: all matter consists of circlons, and all physical phenomena have their roots in the geometry and motion of circlon-shaped forms. Or as he put it in *The Circlon Atom*, "There is nothing in the entire universe which is not made of circlons."

That Jim was aware of the reaction all this was likely to provoke among mainstream physicists was signaled by the name he chose to represent his system: He called it bluntly the Fieldless Universal Circlon Theory, or "the FUCT explanation of reality."

Chapter Seven

SMOKE RINGS

On Catalina, Jim had come across a curious instance of the circlon form, wherein skilled scuba divers would blow circular-shaped bubbles underwater. Aerodynamically, the process is the same as smoke rings. Jim had learned about the possibility of bubble rings from the work of diver-cinematographer Al Giddings, who would direct the underwater photography for James Cameron's epic *The Abyss*, and he determined to learn the art for himself. In *The Cosmic Ring*, he reproduced a grainy black-and-white photograph of Giddings accompanied by a huge underwater ring that hovers in front of him like an eel. Thin and sinuous and quivering with vitality, it seems almost alive, a marine exotic that, like the creature in Cameron's film, appears to have been conjured from the water.

Divers aren't the only ring-making mammals. Dolphins do it, too, and mother dolphins who have acquired the skill will often pass it on to their offspring. Dolphins actually use a different technique from that practiced by divers: Instead of blowing rings directly, they create a vortex in the water with their fins or tails, then they shoot a bubble of air into the whorl with their blowholes. As the water swirls, it spins the nascent bubble into a circle. Dolphins in captivity have been observed playing with such rings, shepherding them through the water with nudges of their snouts and cutting them in two with their beaks. Dolphins have even been observed drawing bubble rings into a helix, creating shimmering spirals in their pools.

As toroidal structures, bubble rings have the same form Jim was proposing for subatomic particles, and he began to wonder if he might use these rings to study how circlon-shaped particles interact with one another. In Jim's theory, circlons link together in networks. Was it possible he could get bubble rings to do the same? In *The Cosmic Ring* he had written that "I have measured photos of air rings, and I have yet to find one that varies in the slightest from . . . the proportions circlons should have." *Bubblons*, as he called them, seemed to offer a clue as to how he might proceed with a practically applied circlon science. If he could make ring-shaped bubbles on cue, his theories would become amenable to empirical test, and Jim speculated that with bubblons he might be able to engage in a new kind of experimental nuclear physics.

The ocean, however, is an inconvenient laboratory, and soon it became one that Jim no longer had access to on a daily basis. In 1976, he and Linda purchased the Green River Gorge property and moved their little family—now with two young sons, Paul and Eric—to the wet wilds of Washington, where they set about building the trailer park. By the time he relocated, Jim had the backbone of his circlon theory figured out. But like all contenders for a major worldview, his idea needed to be fleshed out with details, and for the next thirty years he would pursue his physics wherever and whenever he could make the time. Some years he worked on the theory exhaustively; other years, the demands of the trailer park predominated and he would leave aside formal work to let his ideas simmer in the back of his mind. Always physics was his companion, the background to his consciousness and the fiber of his being.

By 1993, Jim had amassed enough material that he was ready to write a book amalgamating his ideas into one complete theory. Herein he would unite his circlon theory of matter and the new theory of gravity he had been mulling over ever since his road trip in 1962. He would call the book *The Other Theory of Physics*, a title that signaled simply his attempt to offer an alternative to the "standard model" of physics. By November of '93 a first edition was ready, and it was at this point that Jim sent out the package I would receive announcing the publication of his revolutionary work.

On Catalina, Jim had come to believe that bubble rings offered a way forward for experimental investigation of his circlon ideas, but now that he was living in a forest miles away from the sea he realized that smoke rings would be a more effective tool. In 1999, inspired in part by the fact that my husband and I had decided to make a film about his life, he set himself the task of learning how to make giant smoke rings. It was a quest that occupied him for a good part of the following year, during which we would visit him on half a dozen occasions to record his progress and document the birth of what he hoped would be a new era in nuclear science.

Jim had encountered smoke rings through his accidental discovery with soda-bottle bombs, and he knew that process was ad hoc—only about half the bombs produced rings. To study circlons, he needed a method that would produce the forms on cue. And so it was that one evening in the early spring of 1999, I sat with him in his living room in Enumclaw with a couple of empty soda bottles and a pack of cigarettes.

The weather that year was mild, and in the Cascade Range to the east runoff from the winter rain was sending water surging through the gorge. Throughout the trailer park you could hear the sound of its rush, and from Jim's living room, which backs directly onto the gorge, the noise was all-enveloping. The Carters' living room itself is a curious piece of work. It is part of a house Jim designed and built with his own hands, forgoing architects, contractors, and advice. In typical Jim fashion, he had refused to draw up plans, opting instead for an "organic" strategy that infuriated his son Paul—now a grown young man—and pushed even Linda's patience to the brink. Jim has been known to express the views that "walls should be a little askew" and that "if something is perfectly straight, it isn't quite right"—which seems in this case to have been a perfectly realized philosophy. No vertical in the house is plumb. No horizontal is straight. No corner is right. Yet if building norms have been flouted in the Carter residence, it is deviations from domestic norms that first impose themselves on a visitor's consciousness.

Anchoring the space of the living room, a tree trunk sprouts from the floor. Around the trunk spirals a wooden staircase hand-hewn and roughly made. The staircase leads to a balcony and takes one up to the bedrooms on the second floor, beyond which a ladder takes you onto the roof. When I first visited the house, there was grass growing on the roof, giving it a kind of Hobbity feel, but Linda got tired of pulling out the weeds and Jim abandoned the grass. The dirt it was growing in remains, however, and Jim is surely correct that it provides excellent insulation. Someday, he says, he might redo the grass. Or perhaps he'll put in flowers. Other times he proposes building a greenhouse up there as a nice place for Linda to do her exercises.

The house itself is an A-frame construction, with an intimate, cozy sensibility. In the living room, the ceiling is paneled with seven different kinds of wood representing the seven kinds of trees found on the Carters' land, and all of the lumber was milled from trees blown over in storms. The living room is conjoined with the kitchen, and a benchtop counter separates the space so that while Jim is puttering in the living room and Linda is cooking meals, the two of them can be together. This nightly quiet communion is an important part of their lives. On the living room side of the counter, Jim has installed a phalanx of drawers to hold things like balls of string, and bits of wire, and rubber bands, and myriad other items he likes to keep on hand. In the Carter household, "recycling" has always been simply the way you live.

The back of the Carters' living room faces onto the gorge, and from a balcony that abuts it the view is spectacular, a 180-degree panorama of cliffs and forest and sky with the river thundering below. On the evening Jim began his smoke ring experiments, the sound of the waterfall was peaking and the season looked promising for the white-water rafters who would soon invade the area. While Jim played scientist in the living room, Linda began cooking dinner. The menu that evening was elk chili, which in many situations might have been taken as a delicacy. In this case, there been a lot of elk recently, and everyone was frankly tired of elk. The previous winter, Jim had shot an elk in the front yard and the freezer on the back porch had been filled ever since with elk steaks and elk sausages and elk mince. There were undoubtedly more elk meals to come.

It wasn't that Jim had wanted to shoot the elk—he had tried hard not to—but after two seasons in which the great pronged beast had eaten a swath through Linda's vegetable patch, he had given in to the general hue and cry and loaded a rifle. It would have been easy enough to find someone else to do the job—there was an eager line of volunteers among the tenants—but as it was Jim's front yard and Linda's vegetables, it seemed only right that the man for the job would be Jim. Never one given to waste, he had had the carcass professionally butchered, and two hundred pounds of its meat was carried home in neat little paper packages. Jim had taken the hide to a taxidermist to be tanned. Just what he was going to do with an elk hide was hard to say, but as he noted on the evening of the smoke ring experiments: "You never know when a nice piece of elk leather might come in handy."

Jim's quest for a reliable source of smoke rings began that evening with the same technique cigar smokers use: a short exhalation followed quickly by a sharp inhalation. The apparatus in this case was not his mouth, but a soda bottle. Jim isn't a smoker, yet in the name of science he dragged on a cigarette and blew a cloud of smoke into the bottle, where it filled the plastic cavity with diaphanous swirls. The key to a good smoke ring, he explained, is to puff the smoke out the top of the bottle with a short squeeze, then induce a quick retraction to create an immediate in-suck of air. Cigar smokers learn this by feel, using their cheeks as a bellows; Jim mimicked the action by squeezing firmly on the sides of the bottle and quickly letting go. At the clench of his fist, a little cloud of smoke came puffing out the top, tentative at first and only vaguely ring-shaped, but as it sailed into the air it began to pull itself into being, becoming tighter and denser and more clearly defined as it floated across the room.

Jim clenched his fist again, and out puffed a second ring. This time he did it gently so the cloud came out slowly and we could observe the ring forming. Looking closely, I could see the currents of air circling around the main body of the tube, squeezing and stabilizing its form into a tighter, more focused torus.

Jim shot out a faster ring, followed immediately by another, and we watched as the pair of rings sailed toward the ceiling. It was mesmerizing the way their structure seemed to be self-generating; I felt as if I were witnessing the birth (and death) of living things. Their brief lives ended when they crashed into the ceiling and fluttered out of existence like wounded butterflies.

Figure 10. Hermann von Helmholtz (AIP Emilio Segrè Visual Archives)

Jim did not know it then, but the process of smoke ring formation had been studied extensively in the nineteenth century by the German physicist Hermann von Helmholtz. Helmholtz had shown that the self-generating effect resulted from the innate forces of fluid dynamics that act to reinforce the emerging tubular form. In a paper published in 1858, Helmholtz had calculated the flow around a nascent "vortex ring" and had proved mathematically that once a ring begins to form, the natural dynamics of the surrounding air will act to pull it more tightly into being. Helmholtz's equations suggested that it should be fairly easy to make smoke rings, and on a small scale it is; but practically speaking, as the size increases the physical challenges compound. Jim dreamed about making rings at least a yard in diameter, on a scale similar to the bubble rings he had blown underwater, and it wasn't easy to see how such large ones could be made. While it is one thing to make smoke rings on the scale of a few inches, it is quite another to make them several yards across. Yet Jim knew that large rings were possible because volcanoes sometimes produce them. On Mount Etna in Sicily, the Bocca Nuova crater puffs out smoke rings up to two hundred meters in diameter. A spectacular display of Etna rings would occur the following year, in 2000, when between April and July the volcano produced up to one hundred rings a day. Some of these aerial circlons would rise to an altitude of five kilometers and last as long as seventeen minutes.

To my mind, Jim's first attempt at giant smoke rings didn't look like a promising setup; the equipment in this case consisted of a dog bowl and a sack of gunpowder. Jim had set up the apparatus in his front yard, an acre or so of land where Linda has her vegetable patch and where he keeps his cars. When I first started visiting the gorge, the cars would be parked everywhere, but these days Linda confines them to an area by the side of the house, and the spacious lawn is now clear of automotive debris. Screened from the road by a barrier of Douglas firs and redwoods, this is the laboratory where Jim tests his ideas.

On the morning in question, Jim had set up the dog bowl on the lawn, and the dogs—who could not be dissuaded from an interest in its contents—had been shut inside the house. Rebar, the Carters' labrador cross, could be heard from within howling with doggy outrage. The bowl itself was one of those canonical models with a ring-shaped channel surrounding a central hump. Jim had filled the channel with gunpowder, and his plan was to ignite the powder in the hope that its circular shape would be replicated in the resulting cloud of smoke. If this didn't work, he reasoned that at least he would have ruled out one option.

Had Jim read up on the literature of "vortex ring" generation, he would have known in advance that this was doomed to fail. Helmholtz had shown that the process of exhalation followed by inhalation was aerodynamically necessary for smoke rings to form. But research would be antithetical to Jim's ethos. "If you want something done, you do it yourself," he likes to say, and if you don't know how, you figure out a way to teach yourself. In Jim's world, life is an activity to be learned by experience, not by books. In this, he takes his cue from the seventeenth-century philosopher Francis Bacon, the man who first articulated the "scientific method." As an antidote to the ages-old practice of taking on faith what the Greek philosophers had said, Bacon insisted that science should be conducted through personal observation. Throughout the Middle Ages and into the Renaissance, most of what had been said about the functioning of nature had been a rehashing of Aristotelian ideas. Bacon advocated a new way of exploring the world in which direct engagement was key. It was the birth of what we now take for granted as *experimental* science.

Throughout Bacon's writing, he insisted that those who wished to understand the material world must test it and measure it themselves. If that now sounds like a mundane insight, it is well to remember that in the early seventeenth century it was a genuinely novel thought. Jim's allegiance to Baconian-style engagement is one of the things he points to as a mark of a true scientist. Indeed, one of his beefs with mainstream theoretical physicists is that despite their avowals of Baconian principles, too many of them have abandoned his commitments in their work. A blackboard and a supercomputer are the tools of the trade these days, and speculation unchecked by empirical test is once again rife in academic circles. String theory, brane theory, and hyperspace theory: None of these are supported by data, and none would pass muster in Bacon's terms.

And so it was that on a clear Washington morning, Jim Carter confronted a dog bowl full of gunpowder.

Out at the gorge, where most men are hunters and there is a rifle range in earshot, the sound of an explosion is unlikely to alarm the neighbors. Nonetheless, to my eyes the quantity of powder in the bowl seemed sufficient to disturb a fair bit of peace. Like most urbanites, I suppose, I had gleaned my knowledge of gunpowder from watching westerns. Actual explosions were a special effect remote from my experience, and I had never fired a gun in my life. I assumed, naively, that the more powder you exploded, the greater the bang would be, and I was somewhat alarmed at the amount Jim was pouring into the bowl. To ignite it, he trailed a fuse across the yard, and as it sizzled across the lawn I braced myself for a monumental Bang. The shock instead was the lack of drama, for the powder erupted with a velvety "pouf," as if someone had stuffed a pillow over the source. I learned that day that the sound of a gun results not from ignition, but from the confinement of the explosion in the chamber. Score one for Bacon. Zero for my education.

Accompanying the "pouf" was a cloud that ascended above the bowl with what at first looked like a promising ring structure. But as it rose into the air the circle exploded, resulting in a mushroom-shaped cloud like a miniature atomic bomb. Jim tried again, and again the ring exploded. Again and again each attempt erupted in a Bikini Atoll cloud. Had Jim's supply of powder not been finite, we could have gone on forever for the sheer dreamy pleasure of this private Manhattan Project—the little nuclear clouds were beautiful to watch, and you couldn't deny we were learning something from the dog bowl. In the open air, a circular cloud of smoke simply couldn't sustain itself enough to form a ring.

Given the experience with the dog bowl, Jim now reasoned that the explosion should occur inside a pipe; and always one to use materials at hand, he dragged out a massive piece of iron pipe from a nearby shed and bolted it to a fence post at the end of the drive. The sound nearly burst my eardrums and must have been heard by everyone in Enumclaw. This time, however, there wasn't even a hint of a smoke ring. Subsequent blasts yielded no better result. Undeterred, Jim tried different gauges of pipe—eight-inch, four-inch, six-inch—any old tube he could lay his hands on. The explosions were deafening, the rings elusive. Jim tried covering one pipe with a metal cap, and I feared the whole setup would explode. The dogs went berserk and the boom registered on the Richter scale, but nobody got hurt and in the name of science Jim was having a ball. After three days, the gunpowder sack was empty and we hadn't got any closer to a decent ring. Jim was elated. He had narrowed the field of possibility by several options and was taking it all in his stride as "a learning curve." By my next visit, he declared, he'd have smoke rings down pat.

Once Jim conceived of circlons, the idea proved to have power beyond its original purpose. Jim had come to circlons as a way to explain the structure of matter, yet on Catalina he also came to believe he could use them to explain electromagnetic energy. Einstein's equation $E = mc^2$ had famously shown that matter could be transformed into energy and vice versa. If matter consisted of circlon-shaped particles, then it stood to reason that electromagnetic energy must be something similar. Jim's hope for a synthesis between the two was something he shared with the physics mainstream, and the difference between matter and energy remains one of the greatest open questions in contemporary theoretical science. Einstein's equation had shown *quantitatively* how matter is related to energy—it tells us precisely how much energy is released when matter is annihilated, but the equation tells us nothing about the *qualitative* difference between the two. When matter gets transformed into energy, *what* actually changes? After more than a century of thinking about this question, physicists don't really have the answer. Many hope that a new generation of particle accelerators will help to shed some insight.

The current best bet from the mainstream involves a particle known as the Higgs boson, which is sometimes called "the god particle." The Higgs boson was a primary reason why the United States started to build the Superconducting Super Collider in the early 1990s. The machine proved so expensive that Congress opted to discontinue the project, deeming that even a "god particle" could not justify a $13 billion price tag. Detecting Higgs bosons is now a priority for the Large Hadron Collider at the CERN facility in Europe, currently the most powerful accelerator on earth. No one has ever seen a Higgs boson and some physicists are beginning to doubt that it exists, but if it doesn't exist, then physicists' grasp on the difference between matter and energy is even more tenuous.

Whatever matter *is*, its *behavior* differs from energy in the important respect that particles of matter can stand still. Particles of energy, or "photons," are always in motion, for they are always traveling at the speed of light. What makes matter "matter," and the reason it *can* stand still, is that particles of matter possess what physicists call a "rest mass." Photons don't have a rest mass, so they can't stand still. Without an inherent massiness, they are condemned to perpetual flight. According to the special theory of relativity, when energy is converted into matter, rest mass is created. Conversely, when matter is converted into energy, rest mass is destroyed. In some sense, then, matter and energy are similar, but they are also very different, and the precise nature of the difference remains mysterious. The Higgs boson is supposed to be a bridge between them, however its no-show in accelerators is becoming rather troubling.

When physics students at college learn about matter-energy transformations, they are told about gamma rays, and in *The*

Cosmic Ring Jim also picks up the gamma ray story. According to mainstream physics, when an electron and positron particle collide they annihilate each other, and in their place two gamma ray photons are created. Nuclear explosions generate huge bursts of gamma rays, and U.S. spy satellites search for such bursts as proof of foreign nuclear tests. In *The Cosmic Ring*, Jim discusses gamma ray production as an introduction to his own views. Specifically, he rejects Einstein's idea that rest mass is annihilated. Instead, he says, when an electron and positron collide, the circlons that make up the two particles get reconfigured so that the slow-moving "matter" circlons are transformed into high-speed "gamma" circlons. According to Jim, the difference between matter and energy is that both are different states of the underlying springy stuff that is the basis of everything. As with so much else in contemporary physics, Jim sees the "rest mass" problem as an artifact of mathematical models dreamed up by theorists who have abandoned Baconian ideals.

By the fall of 1999, when my husband and I visited the gorge again, Jim had made progress with his smoke rings. He had actually been doing research and had come across a suggestion that manifested itself in an innovative use of Saran wrap. Again in his living room he demonstrated with a soda bottle. We were, now, in the post-elk era. This time Jim had cut the top off the bottle, and across the opening he had stretched a piece of Saran wrap. Now when the bottle was full of smoke the Saran wrap acted as a membrane and you could tap on its skin like a drum. In the side of the bottle Jim had cut a small circular hole, where the smoke rings would come out. The process of making rings was now a lot more efficient because the Saran wrap had bounce, making the exhalation-inhalation dynamic more pronounced. Though Jim didn't know it, something like this is exactly what Helmholtz had advised. A series of taps on the Saran wrap now produced a cascading series of rings: tap, puff, tap, puff, tap, puff. Soon the air of the living room was filled with rings like a flock of little phantoms. The harder Jim tapped on the Saran wrap, the faster the smoke rings flew, and with a bit of practice he was shooting them across the room like bullets. For a while he amused himself by shooting them at Rebar, who jumped and barked in a frenzy until Linda told everybody to "stop carrying on." With really gentle taps Jim could make rings that almost hovered in the air, which was great for close-up inspection. When you watched carefully, you could see the currents of air swirling around the tube, pulling it tighter into being.

The Saran wrap gave Jim the information he needed to scale up the experiment to the size he desired—it was the *membrane* that provided the action/reaction dynamic critical to the formation of good vortex rings. I am not sure what I expected on my next visit to the gorge, but certainly not the behemoth he'd installed in the yard. He had bought an old oil storage tank, ten feet long and six feet in diameter weighing several tons, which he had hauled home from some scrap yard. There it sat on the front lawn like a gargantuan toy rigged up with a membrane of rubberized sheeting stretched across one end. Jim had purchased the tank to make a hot-water boiler for the house, but when he realized it would make an ideal smoke ring generator, he had no hesitation in changing plans. On one end of the oil tank was the piece of rubber sheeting, and at the other end was an open porthole, so the whole thing constituted a vast scaled-up version of the soda-bottle experiment. To scale up the tap of his hand on the Saran wrap, Jim had attached a huge iron grommet to the center of the rubber sheet, like a ring in the nose of a gigantic pig. Through the ring he had threaded a rope that was slung over the branch of a nearby tree. By hauling on the rope, he could drag back the rubber sheet so that when he let it go it would, hopefully, deliver a powerful enough thwack to propel the volume of air within.

The one question remaining was how to fill this enormous vessel with smoke. Jim had an answer to that as well, and so it was that the next day we drove into Seattle to a disco and party rental store to rent a smoke machine. "Authentic Night Club Feel!" the brochure promised: "Dazzle Your Guests!" "Thrill Your Friends!" "Keep Out of Reach of Children." Smoke machines work by dripping oil onto a hot plate, so we also purchased three bottles of premier smoke fluid. The smoke it produces is thick and white and perfect for making rings.

By the time we got back to the gorge, a little crowd had turned out to watch. By now it had been six months since the dog bowl episode, and expectations were running high. Everyone in the trailer park knew something was afoot. Linda and Eric were there along with several of Jim's workers, including Richard, a brilliant auto mechanic who helped to keep the fleet of cars running. Among the spectators also was Melody, owner of the campground's resident Airstream trailer. Melody is one of the Carters' longest-term tenants and had been living at the gorge for more than twenty years. Born with a genetic disorder, she smoked constantly to distract herself from the pain of joints that were being destroyed by a cruel form of arthritis. Everyone had their dogs with them, and of course Rebar was on hand.

While Jim is a man perfectly happy to be doing things for

himself, he also has the instinct of a showman and knows how to hold a crowd. While everyone waited he fiddled with the membrane and checked the tension on the rope. He waited for a cloud to pass, and as its shadow crossed the lawn he speculated about the size of the rings the tank might produce. When everything finally seemed ready, he decided the tank needed more smoke and topped it up with the disco machine. By the time the action started, the little crowd itself was about ready to explode.

Finally Jim tugged on the rope and let the massive membrane fly. The thwack it made thudded through the tank, drawing forth a deep bass, metallic-sounding note. An audible pause followed in its wake, long enough for the specter of failure to flitter through my mind. Then, as if in answer to the call of some ancient hunting horn, a huge white smoke ring came surging out the porthole and charged across the yard. Jim dragged on the rope again, and another thwack of the membrane produced an equally perfect twin, swift and meaty, rocketing into the air above our heads toward the firs. The crowd, prone to an ingrained "Gorgie" skepticism, let out noises of approval—a whistle from Richard, a whoop from Brian, furious barking from Rebar. Jim himself was chuffed and allowed himself to bask in the glow of their approbation. For the next hour he delighted dogs and humans alike as he conjured circlons from the mouth of the great steel dragon, transmuting disco fluids into the stuff of atoms in a triumph of junkyard alchemy.

Gamma rays aren't the only kind of energy Jim hoped his circlon theory would explain. In *The Cosmic Ring*, he proposed that all varieties of electromagnetic radiation are circlons moving at the speed of light. Since circlons underlie both matter and energy, he realized that all of nature could be united in a cosmic chain of being, with particles of matter and photons of energy each constituting different levels in a universal hierarchy. Matter could be changed into energy and energy into matter because both were different manifestations of the same underlying stuff. Every kind of energy—gamma rays, X-rays, visible light, radio waves, and so on—would have its own *order* of circlons. The "electromagnetic spectrum" would be nothing less than a sequence of different-sized circlons, and Jim now renamed it the "circlon spectrum."

In the electromagnetic spectrum of mainstream science, energy is divided into seven major types—gamma rays, X-rays, ultraviolet, visible light, infrared, microwaves, and radio waves. Gamma rays have the highest energies and shortest wavelengths; radio waves have the least energy and longest wavelengths. Jim being Jim, he set about naming things in his own way. In his circlon spectrum there are thirteen distinct energy types, each represented by its own class of circlons. Those associated with the gamma rays, Jim named the "gammalons." Those associated with ultraviolet, he called the "violons." The infrared circlons were the "redlons," the X-ray circlons were the "excirlons," the radio waves were "ralons." To these canonical categories he added six additional types, the worlons, the solons, and the bubblons, among them. At each step of this hierarchy "energy" and "matter" are united, so that the class of gammalons, for example, includes not just gamma ray photons, but the associated matter particles, the electrons and positrons. In Jim's theory, when particles "annihilate," the underlying circlon-stuff simply changes from the "matter" state to the electromagnetic wave state. As Jim sees it, nothing is created or destroyed in this process, the springy stuff of the circlons just changes its state of being.

In *The Cosmic Ring*, Jim illustrates each class of circlons with a photograph of a physical object to which it relates. Thus the ex-circlons are represented by an X-ray of a human hand studded with buckshot. The pieces of shot embedded in the limb seem to be a metaphor for the circlons themselves, which according to Jim's theory get captured by our tissues. As the circlons associated with ultraviolet light, violons are the cause of sunburn, while redlons are responsible for temperature change. Ralons, corresponding to radio waves, are vast "football-field-sized" circlons zinging through space.

But what are we to make of "bubblons," a category not singled out by mainstream science? These are circlons a meter or so in diameter, corresponding to electromagnetic frequencies with wavelengths on the meter scale. Mainstream physics recognizes this as part of the shortwave radio band, and Jim notes in the book that bubblons "are the medium by which television is transmitted." This part of the spectrum is also where we find "citizen radio bands" and radar. To mainstream physicists, this part of the spectrum is of no particular theoretical interest, yet such signals possess a quality that has commanded human attention through the ages. As Jim tells us, this is the scale that corresponds to our own body size. Man himself is a bubblon-sized being.

We cannot see circlons directly, any more than we can see electromagnetic waves, but on Catalina Jim began to realize that the world is pregnant with these forms. The circular bubbles that divers blow are nature's way of manifesting bubblons in water. Once Jim became aware of the circlon form, he began to recognize that indeed it manifests at every scale of our world.

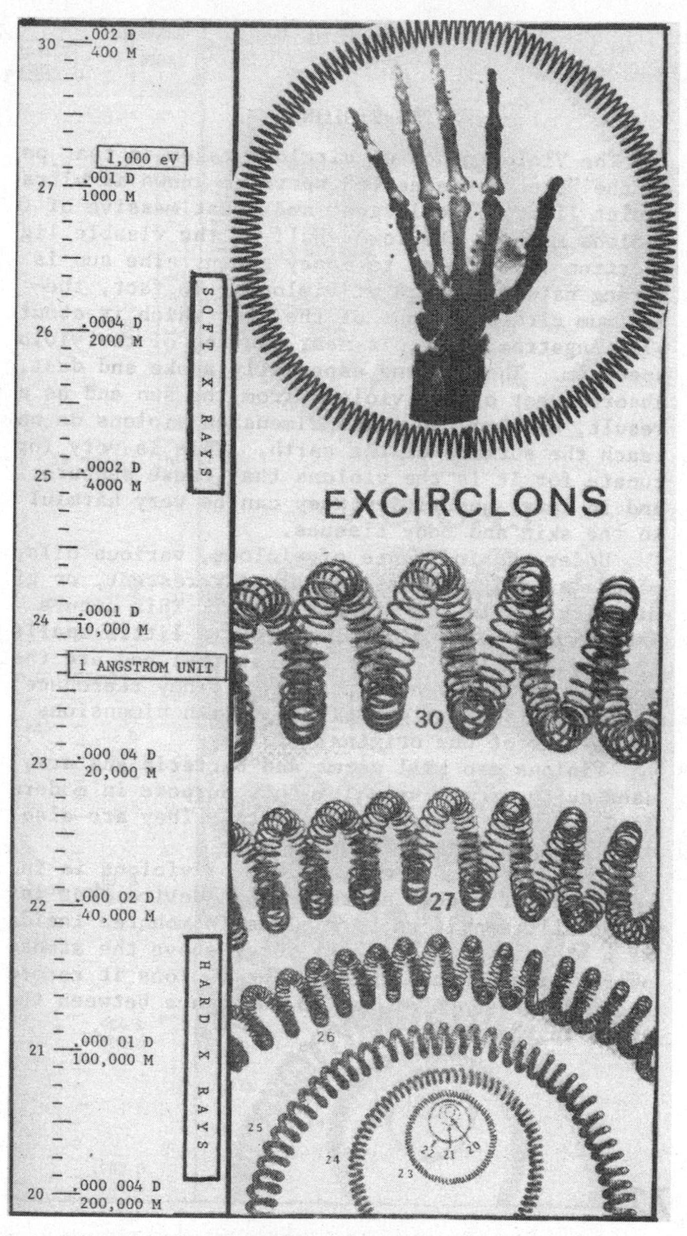

Figure 11. Page from The Cosmic Ring *illustrating the physics of X-rays.* (Jim Carter)

The magnetic field of the earth, for example, wraps our planet in a circlon of terrestrial proportions. Jim calls this a "worlon." Scaling up further still, we find the "solons," witnessed in the great "loop prominences" of plasma that spew out from the surface of the sun. Going further still, in the constellation Lyra we find a giant ring nebula thousands of light-years across, a circlon to awe even Jim.

Back at the Green River Gorge, Jim's oil tank had demonstrated the efficacy of the membrane approach, yet this apparatus was too cumbersome for practical research. Simply filling the tank required half a bottle of disco fluid, and the huge scale of the vessel made it impossible to maneuver. It was only on my final visit to the gorge that year that Jim hit upon the ideal mechanism for smoke rings, a solution that had been under his nose all along. His equipment this time was a garbage can, or more precisely three large black plastic garbage cans of the type you find at campgrounds. Each can was three feet high and two feet in diameter. Again, Jim had covered the tops of the cans with a piece of rubber sheeting and in the side of each can he had cut a hole about eighteen inches across. Again he had rented a disco machine to generate the smoke. He had placed the cans across the yard, in a line, and we waited for an afternoon when the wind was low and the skies clear. Finally, when a cloudless day arrived, Jim filled the cans with smoke and began to beat on the membranes, running from can to can as if he were playing a set of bongo drums. Gently at first, like a rock musician warming up, he patted each rubber skin, and as he slapped its surface a perfect smoke ring came sailing out of its mouth. Harder slaps produced denser rings, and a good solid whack achieved the result he had been striving for all year—superbly formed rings went sailing across the lawn.

Jim aimed one of the cans into the air to see how far the rings would go, and the best ones made it to the end of the yard, two hundred meters away. They faded from view only when the smoke thinned to a point that it was no longer visible against the sky. Miraculously, they retained their structure to the end, "dying" not from loss of form, but from the inevitable slow fade-out as their size increased. Jim's greatest triumph that day was making two rings collide. By carefully aiming cans at an angle to each other and by judicious timing of the slaps, he was able to achieve a sideways swipe. Just as he had imagined on the abalone boat out at Seal Cove almost thirty years before, when the rings collided they bounced apart like billiard balls. Although their shape was disturbed by the interaction, and would be destroyed if the angle of incidence was high, mostly, after a smoky convulsion they found their way back to their circular ideal.

Against the backdrop of the setting sun, Jim made smoke rings until the light faded. As they cavorted in the gloaming, he mused about the application of his DIY apparatus to experimental nuclear science, an activity normally confined to the physics elite with their multibillion-dollar particle accelerators. Jim has always dreamed of having his ideas tested in one of these machines; he has even designed an experiment he believes would distinguish between his theory of matter and the "standard model" of physics. But he is under no illusions about his chances of being welcomed through the door of Fermilab or CERN, whose hugely expensive facilities are controlled by peer-review panels. Jim understands that an outsider from Enumclaw is not going to make that grade. Now with his garbage cans and smoke machine he can test his ideas in his front yard for a fraction of the price. As a particularly long-lived smoke ring sailed through the air, Jim let loose the opinion that "this is just a way I can do nuclear experiments that doesn't take the millions and billions of dollars that the particle accelerators cost." Budget-conscious in his own life, he has found a way to circumvent not only the norms of academic theory, but, perhaps more impressive, its fiduciary constraints. Watching as that smoky Methuselah wafted toward the forest, I was struck by a thought: Like the smoke rings, Jim's physics pulls itself into being. Circlon science, like a circlon of smoke, is a self-generating phenomenon.

Chapter Eight

CREATING THE WORLD

JIM CARTER WASN'T the first person who had dreamed that smoke rings might open a door to the atomic realm. In the middle of the nineteenth century, a curiously similar set of experiments was conducted by a pair of scientists in Scotland driven by very similar motives. Though this steam age science is not often mentioned in official histories, it ushered in a strand of thinking that would eventually lead to string theory, thereby assuming a central role in the landscape of twenty-first-century physics. Jim did not know it, but he was following in the footsteps of scientific giants who were themselves trying to understand what atoms are. Jim's unawareness of this work can hardly be held against him, for most professional physicists are similarly in the dark. So aberrant are these efforts deemed, they are rarely discussed when the story of science is told.

Behind the enterprise were two exemplars of nineteenth-century British intellectual life: Dr. Peter Guthrie Tait, a master mathematician and professor of natural philosophy at the University of Edinburgh, and his lifelong friend Sir William Thomson, a physicist who was already one of the most famous scientists in England and who was on his way to becoming an international celebrity. Tait and Thomson represented the quintessence of science in their day. Thoroughly trained academic superstars, they were part of an advance guard pressing forward to a new era in which physics and mathematics were increasingly conjoined. In smoke rings the two disciplines were intertwined, and through an exploration of these forms they hoped to make progress at the forefront of both fields.

In their personal lives, both men stood larger than life—Tait literally so. At six feet tall, with a barrel chest and an enormous red beard, he radiated vitality. When he wasn't solving some arcane mathematical problem, he could often be found striding across a golf course whacking balls with the gusto of an accom-

Figure 12. Lord Kelvin, Sir William Thomson (AIP Emilio Segrè Visual Archives, Zeleny Collection)

plished athlete. Scotland, of course, was the home of golf, and Tait was so in love with the game that he wrote several research papers analyzing the dynamics of the little pitted ball. His passion for the outdoors was matched by an equal appetite for intellectual life, and there were few subjects to which he did not feel his enormous talents might not positively contribute. The middle of the nineteenth century was a time when a man might still aspire to know everything, and Tait was among the elect few for whom that seemed an attainable goal.

Thomson, his colleague, was at the forefront of science both as a participant and also as an entrepreneur. In 1866, he was knighted by Queen Victoria for his contributions to the laying of the transatlantic telegraph cable, which inaugurated the era of international telecommunications, and during his life he amassed fortune from his innovations in the design of electrical equipment. A professor at the University of Glasgow, he kept one foot in the academic world and one foot in the public sphere, where he loved to work on real-world problems and where he played a prominent role in helping to communicate scientific discoveries to the public. At the dawn of the media age, Thomson understood as few of his colleagues were yet able to grasp the power of the press in marshaling support for science.

Thomson was a pioneer in the science of thermodynamics, the branch of physics that underlies our understanding of heat flow and which thereby helped to enable the vast improvements in steam engines critical to the latter phase of the Industrial Revolution. For this and myriad other contributions, he would later be granted the title of baron, an extraordinarily high honor for any scientist to attain. Accepting the title, he chose the name Baron Kelvin after the river Kelvin that flowed through

Glasgow near his laboratory. His work on the physics of heat led to the revelation that at the lower end of the temperature scale there is a limit: While hotness can be infinite, coldness is subject to a lower bound—there is an "absolute zero," as it were. Thomson's research on this subject led to the development of the absolute temperature scale, whose unit—the degree Kelvin—is named in his honor. The following century, an equally glittering tribute would be paid to him when the Kelvinator refrigerator was christened, thereby making Thomson perhaps the only man in history to become both a fundamental physical unit and a household brand name.

Unlike Jim Carter, Tait and Thomson had become interested in smoke rings through someone else's work. In 1858, Helmholtz had published his paper on the physics of "vortex rings," drawing on the emerging science of fluid dynamics to make predictions about how these forms would behave. Specifically, the mathematics had led him to insights about what could be expected to happen when two smoke rings are traveling close together. Tait was eager to find out if these insights were true—would the mathematical predictions be borne out in the material world, or were they just artifacts of an abstract model? To test Helmholtz's claims, Tait set about constructing an apparatus to make smoke rings on tap. From this humble beginning, he and Thomson would be led to considerations of the utmost kind—Thomson would eventually posit a theory about the nature of matter, while Tait would propose a model of the universe that enabled the possibility of life after death.

This adventure in smoke ring science began quietly on a chill winter's day in 1867 in Tait's rooms at the University of Edinburgh. The world of physics was changing fast as new mathematical techniques opened doors to understanding in ways undreamed of by previous generations. Faraday's fields were being taken up by a generation of physicists who embraced mathematics with gusto. Thomson, in fact, was the first person who had tried to give the field concept a mathematical framework and in 1865, Maxwell had succeeded in that goal. In London, De Morgan was compiling his *Budget* and reconforming the foundations of mathematics itself. Physics was in the midst of a golden age, and fluid dynamics was one of the fields in which new horizons were being explored. Helmholtz's paper, though ostensibly about a modest subject, was at the vanguard of what would turn out to be one of the most important and difficult topics in all of physical science.

Having read Helmholtz's paper, Tait had set out to construct a "vortex ring generator" to put his predictions to the test. The design of his device was the same one Jim would eventually hit upon, but Tait had gotten it right on the first go-round because he had followed Helmholtz's advice. What Tait also had in common with Jim was his use of household castoffs. Instead of a garbage bin, the body of his generator was a wooden packing crate. He too had sawed off one end, over which he had draped a damp towel. This was to be his membrane. In the opposite end of the crate, Tait had cut the requisite circular hole, but as the disco revolution had yet to occur and he did not have the luxury of an off-the-shelf smoke machine, he had had to concoct his own source of smoke, no small task at a time when the science of chemistry was still in a state of formation. (Mendeleev was at this moment working away in Russia on the periodic table.) By the time Thomson arrived in Tait's rooms, all the supplies were in place and the vortex ring generator had been set up on the desk. We know what it looks like, for in one of Tait's books he included a pen-and-ink drawing: There it sits like an eager little dragon angled up on a chock of wood, with its blowhole pointing into the air.

Let us pause for a moment to imagine this scene: We are at a university in Edinburgh. It is winter, and outside the snow is piled deep. Inside, a fire would normally be lit, but the fire isn't burning today because the experiment must be conducted in still air to minimize crosscurrents. The windows are closed, and the book-filled room is hushed and cold. Tait prepares the ingredients for the smoke: ammonia, sulfuric acid, and a handful of salt. He throws the mixture into the box and the resulting gases combine to form a foul-smelling cloud of sal ammoniac. As Thomson watches, Tait whacks the towel with his hand and out of the blowhole shoots a neat little ring that sails toward the bookshelves. A series of whacks produces a series of rings, each about a foot in diameter, thick and meaty like sausages. Thomson, delighted, tries poking one with his finger, setting it wobbling like a jelly. He pokes another more forcefully, causing violent trembles in its form, yet like magic it regains its structure before crashing into the shelves. Marveling at its integrity, Thomson admires the power of the rings to assert themselves against the destructive forces of turbulence. It is as if they are imbued with internal knowledge of what their form *must* be. "Just as if they were solid rings of India rubber!" is how Tait would describe them in his book.

Tait has constructed a second box so that Thomson can play, too, and soon the two men are shooting off rings together like cannoneers. Towels thud, rings burst forth, the room fills with fumes. Out in the corridor, the usual proprieties of academe roll on regardless, while malodorous gases amass in the hall. Tait is eager to see what will happen when several rings collide, and the two men aim their boxes into the air. Though tricky to coordinate, they rejoice when finally two rings graze each other like billiard balls and miraculously bounce apart. As physicists, they are hooked.

The success of the Edinburgh experiments was all Tait could have wished for, and soon Thomson was building boxes of his own in Glasgow. For the next several years, the two scientists exchanged letters reporting on their explorations: What happens when the boxes are different sizes? Or different configurations? Does a cubic box behave differently from a rectangular one? (It doesn't.) What happens when the blowholes are different shapes—does a square hole produce square rings? (No. Whatever the shape of the box or the hole, circular rings result.)

What is the best recipe for making smoke? Tait wrote to his friend that "the true thing is $SO_3 + NaCl$ . . . NO_5 is DANGEROUS," he warned in capital letters. "Have you ever tried plain air in one of your boxes?" he asked. "The effect is very surprising. Put your head into a ring and feel the draught."

It is wonderful to imagine the scenes that must have taken place across this interchange: two bearded, balding beacons of British scientific rectitude nuzzling their whiskered cheeks into the drafts of oncoming smoke rings. "There is no sensation until the vortex ring is almost close," Tait wrote. Then, just as the ring impacts on the face, there is the unexpected sensation of "a sudden blast of wind flowing through the center."

In 1876, Tait reported on these experiments in his book *Lectures on Some Recent Advances in Physical Science*, a populist account of the latest advances in physics aimed at the general public. Based on a series of lectures he had given, the volume covered major developments in this rapidly developing field, including thermodynamics, electromagnetism, field theory, and fluid dynamics. The latter is hardly the stuff of newspaper headlines, yet its impact on our lives could hardly be greater: Airplanes, automobiles, and oil pipelines are all designed using the equations of fluid flow. You may never have heard of the Navier-Stokes equations, but twenty-first-century living would be unthinkable without them. They are used to model the weather, ocean currents, the motions of stars in galaxies, the movement of blood through veins, and the flow of liquids in pipes, making them an essential tool for factory design. These are the equations that also underlie smoke rings and the mathematical foundation Helmholtz had used in his analysis of these forms.

Helmholtz's paper had drawn on the Navier-Stokes equations to consider how a vortex ring would behave in what is known as an "ideal fluid," the platonic or mathematical ideal of a fluid that may also be thought of as an infinitely fine gas. (For the purposes of Navier-Stokes, gases are regarded as fluids and the science of "fluid dynamics" covers both states of matter, hence planes flying through air are governed by these equations along with boats traveling through water.) Using Navier-Stokes, Helmholtz showed that once a vortex ring begins to form in an ideal fluid, it will last forever; the motion of the fluid around the ring becomes self-generating. No *actual* fluid can sustain a ring forever, but Helmholtz's analysis revealed why real smoke rings floating through real air should be inherently stable. As a by-product of his analysis, Helmholtz also found that the mathematics predicted something extraordinary about the interaction of two rings. If two rings were traveling in the same direction coaxially (with their centers aligned), then the mathematics declared that the natural dynamics should cause them to engage in a kind of leapfrogging dance.

To see how this works, imagine two smoke rings traveling together about a foot apart; they are the same size and traveling at the same speed, one behind the other. All our intuition tells us they should remain this way, one behind the other like two rolling balls. But that is not what the equations predict. According to Helmholtz's analysis, the intrinsic dynamics should cause the leading smoke ring to enlarge and slow down, while the trailing ring should shrink and speed up. The equations predict that in fact the trailing ring should eventually pass through the leading ring and replace it in the first position—whereupon the roles should reverse and the whole process repeat itself so that the two rings alternately "penetrate one another" in a skipping kind of dance. The "pursuer" should become the "pursued," and then vice versa, is how Tait would describe it in his *Lectures.* This was a pretty extraordinary prediction and wholly counterintuitive. Tait determined to put it to the test.

Coordinating two smoke rings in a line was tricky to achieve in practice, but soon Tait and Thomson had verified Helmholtz's prediction. The best method was to fire off one ring, wait a beat, then fire off another one slightly harder. As the second ring approached from behind, the first indeed began to slow down and simultaneously enlarge, while the second one shrank and speeded up, enabling it to pass through the first. Tait and Thomson were able to demonstrate the leapfrog game for several of these skips, and soon Thomson was writing triumphantly to Helmholtz to report on their success.

Two vortex rings moving in the same direction play a game of leapfrog; even more remarkable is the behavior of two rings traveling *toward* each other. Here we naturally expect them to collide and annihilate, but, again, this is not what the equations predict. According to Helmholtz's analysis, as two rings approach each other from opposite directions, they should begin to draw off each other's energy. As a result, the equations hold that *both* rings will begin to expand and slow down. Helmholtz predicted that indeed they would slow down so much, they would never actually collide. Against all the preconceptions of our sensory experience, the mathematics predicts that two approaching smoke rings will never physically touch, they should just continue to get bigger and slower, as if time itself were stretching around them. This was a truly bizarre prediction, and Tait determined to test it as well. Here again, Helmholtz was vindicated. Even Tait's enormous confidence in the power of mathematics could not blunt his surprise at seeing two smoke rings engage in this near mystical pas de deux.

Following Tait and Thomson's experiments, their friend James Clerk Maxwell joined the fray. A mathematical genius, Maxwell now analyzed the equations for the three-smoke-ring case and showed that when all three were traveling in the same direction, they should *all* engage in the leapfrogging game: **A** would skip through **B**, then **B** through **C**, then **C** through **A**, and so on. Sadly, such a trick was too difficult mechanically to pull off with real smoke rings, but Tait and Thomson did not have to content themselves with purely mental pictures. Maxwell drew a series of diagrams depicting the three-ring encounter and mounted them in a zoetrope, a pre-cinema viewing device that enabled people to experience the illusion of motion by watching a spinning sequence of stills. As viewers looked through a slit in the side of the zoetrope, they could watch an animation of three rings eternally passing through one another like puppies playing a game of tag. This enchanting example of steam age virtual reality is preserved today in the Cavendish Laboratory in Cambridge, where Maxwell served for many years as director.

Once Tait and Thomson had verified Helmholtz's predictions, they could have allowed the subject to rest. Neither was lacking for things to do. The foundations of physics were shifting, and both were at the forefront of this movement. Thomson, for a start, was helping to formulate a mathematical description of heat and energy and to wonder what makes stars shine. Yet over the years smoke rings continued to exercise a magnetic pull on his mind. As he watched rings sailing through the air, he began to ask himself if they might be a model for atoms. Might matter itself be made up of minute vortex rings? Thomson was a physicist, and as such he knew that atoms had to exist for a very long time. Wasn't that just what Helmholtz had shown vortex rings would do in an ideal fluid? Perhaps, Thomson speculated, atoms were made up of little vortex rings floating about in some superfine ethereal substance. Science even provided the perfect candidate, the so-called luminiferous ether, a near ideal fluid that nineteenth-century physicists believed to be permeating space. This ether would be a perfect medium in which "vortex atoms" might form.

Thomson reasoned that some atoms might be composed of a single vortex ring. Others might be made from several rings linked together. He imagined vortex rings combining into "daisy chains" and groupings with three or more rings "running through one another." "Every variety of combinations might exist," he rhapsodized to Helmholtz. If the idea proved correct, it would resolve one of the oldest questions in science. Just as Jim Carter would do a century and a half later, Thomson came to believe that vortex rings might explain how different kinds of matter have different physical properties. Why are some materials hard and others soft? Why are some transparent and others opaque? Thomson imagined all this being explained by the pattern of their vortex rings. In the new atomic science he foresaw, every element in the periodic table would have characteristics that resulted from the assembly of its vortex rings. The more he thought about the idea, the more Thomson liked it: It was simple, it was powerful, it was elegant. How could it *not* be true?

William Thomson and Jim Carter have more than a theory of matter in common. In many ways, Thomson is the model of a scientist that Jim is seeking to follow. Not that Jim is trying to emulate anyone, but if there is anyone in the history of physics to whom he might be compared, Lord Kelvin is a better candidate than most. Like Jim, Thomson was a man with a prodigious imaginative streak—too prodigious, according to some of his more conservative colleagues. He too was a passionate theorizer, a speculator, and a world builder. Throughout his life, Thomson collected bits and pieces of other people's theories and stitched them together in novel ways. As with Jim, this bricolage was apt to produce chimeras that often offended more categorical minds. With his penchant for picking up intellectual baubles, his biographer David Lindley has described him as "a magpie." "Dullness does not exist in science," Thomson declared, and he lived his life as if by some enchanted decree, tinkering and playing like a perpetually brilliant child. Operating as much outside the academic sphere as within it, he loved to get his hands "dirty" with real-world problem solving, especially anything of an electromechanical nature.

Thomson had in fact been a very brilliant child. As a youngster he used to accompany his father—also a professor of mathematics—to classes at the University of Glasgow, and he had a habit of solving problems that Papa's students found hard. At sixteen he published his first research paper, and by the time he graduated from Cambridge University, he had amassed a publication record that most professional scientists would envy. At twenty-two he was appointed to a professorship at Glasgow University himself, thereby becoming his father's peer.

Thomson believed there was no corner of nature that could not be illuminated by clear thinking bolstered by the tools of mathematics. In Lindley's brimming biography, it is hard to believe one lifetime could have encompassed so much. In the field of thermodynamics, he helped formulate the principle we now know as the "conservation of energy"—the idea that energy cannot be created or destroyed. He was a leader in the development of the kinetic theory of gases and the first to give a mathematical formulation to the concept of electric fields. He thought about what made the sun glow, and tried to calculate the age of the earth from basic physical principles, which got him into an enormous battle with Darwinian evolutionists because he argued that the age of our planet was just a few million years—not nearly long enough for the huge epochs of time they said were needed for new branches on the tree of life to form. Thomson wasn't against evolution, he was just committed to physics, and as far as he could tell from the physics then known, the earth could not be more than ten or twenty million years old.

Thomson lived in a world that validated rigor, yet throughout his life he retained a sense of play. As an elderly gentleman, he liked to visit his friend George Gabriel Stokes—he of the Navier-Stokes equations—and the two physicists were wont to discuss scientific matters over breakfast. Their eggs were boiled in an egg boiler on the table, but for Thomson, who was by then Lord Kelvin and a peer of the realm, any old egg-boiling method would not do. The younger physicist Joseph Lamour described the egg ritual thusly: "Lord Kelvin would wish to boil them by mathematical rule and economy of fuel, with preliminary measurements in the millimeter scale." He was applying thermodynamics to his breakfast.

Thomson's brilliance was recognized at the highest echelons of science, and three times he was offered the directorship of the Cavendish Laboratory. Each time he declined. Not wishing to be confined to academic research, he was always looking for ways to make the world better in a practical sense. A story was told in his family that illustrates the point: Once while staying at the family's country house, he was annoyed by the sound of a dripping tap. Unable simply to make a minor repair, he set out to invent a new kind of plumbing fixture, for which he was granted a patent. The same spirit was at work with his activities around the transatlantic cable. While the cable was being laid, Thomson invented the mirror galvanometer, which enabled technicians to measure electrical signals coming down the wire. The galvanometer was a huge success, but readings had to be taken down by hand, and as more telegraph and telephone cables were laid around the world, meter reading became a costly impediment. Thomson longed to automate the process. To that end, he invented a device to record the measurements on paper tape. The prototype worked well enough, but he was never able to make it sufficiently reliable for commercial operation. For that we may forgive him, Lindley says; he had effectively invented the ink-jet printer 120 years before the personal computer revolution.

If Thomson's energies were inexhaustible, his imagination was more so. He almost could not stop himself from inventing new theories. If ever there was a "discoverer," it was William Thomson. Thinking about the sun, he decided he could model the lifetime of a star as a giant liquid drop, an insight that has proved fruitful ever since. In his work on electric fields, he theorized that we could understand the space between charged objects as "a jelly." To a mind as agile as this, "vortex atoms" were not such a stretch, although it is worth remembering that in the middle of the nineteenth century no one even knew yet if atoms actually existed.

Once Thomson had the idea of vortex atoms in his mind, he could not help seeing further elaborations, and just as Jim Carter would do with circlons, so Thomson began to embellish on his original idea. Simple rings, like the ones produced by Tait's boxes, are the most straightforward example of a whole class of objects that mathematicians call "knots." Thomson now speculated that each kind of atom might be a different kind of "vortex knot." While simple atoms might be composed from simple rings, more complex atoms might be formed from more complicated knotty structures. As any knitter knows, knots come in an endlessly frustrating variety, and the subject was one that mathematicians were then beginning to formalize.

What are the properties that make one knot different from another? they asked. How can we tell if two knots are the same? How many knot types are there? It was a bit like cataloging fossils.

Mathematicians of the late nineteenth century wanted to produce a taxonomy of knots just as the naturalists were busy doing with fossils. While Thomson applied himself to the physics of explaining atoms as vortex knots, Tait threw himself into the taxonomic project. They both reasoned that if a taxonomy of knots could be produced, it might also serve as a classification scheme for atoms.

It was here that Tait made one of his lasting contributions to the history of ideas. He became in effect the Linnaeus of knots, producing the first "knot tables." Inspecting each knot creature carefully, he noted the details of its morphology, and arranged them into family groupings that shared common features. He drew sketches, creating an Audubon guide for the species. It turns out there are an infinite variety of knot types, far more than the one hundred or so elements in the periodic table. If atoms actually *were* vortex knots, as Tait and Thomson proposed, then nature had lots of possibilities to choose from.[1]

Sadly, Thomson's hopes for vortex knot theory didn't pan out. The mathematics of the Navier-Stokes equations for knotted structures was too hard to resolve. In Tait's *Lectures*, a rare note of frustration creeps in. So complex is the problem, he says, that it may require "the lifetimes, for the next two or three generations, of the best mathematicians in Europe" to sort out the difficulties. He and Thomson were so enamored with vortex knots, they seriously imagined mathematicians devoting themselves to the task for the next hundred years. When that didn't seem to be happening, Thomson tried to reimagine the theory in a more tractable mode, at one point proposing that our universe is a giant "vortex sponge."

Tait imagined something even more far-reaching, for in vortex atom theory he discerned the possibility for life after death. He presented his ideas on this subject in a book he published in 1875 called *The Unseen Universe*, one of the earliest attempts to give a scientific account of how our universe began. Although the Greeks had believed that atoms are eternal, the new science of thermodynamics made it clear that no physical object can exist forever. If atoms are real, then they too must be ruled by the laws of thermodynamics and therefore subject to creation and decay. Real atoms, like real people, begin in real physical processes. What are these processes? Tait proposed that vortex atoms offer a way forward. If atoms are knots or rings in some kind of ether, then, like real smoke rings, they must come into being at a finite time. In other words, there must have been a time *before* atoms existed when the ether stood alone, unformed, as it were, the way the smoke was unformed in Tait's boxes before any rings were made.

But if the ether exists *before* the atoms, then that raised the question of what the ether itself is made of. Here Tait made a truly inspired suggestion: He proposed that the ether was in turn composed of even smaller vortex atoms, formed in a layer of even *finer* stuff. The lifetime of these superethereal vortex atoms would be a great deal longer than the atoms we encounter but still finite.

These superatoms would in turn be composed of even finer, more long-lived, vortex atoms existing in an even finer ether. And so on it would go, with ever more layers of ether composed of ever finer vortex atoms, each constituting another, more enduring layer of being. All these hidden layers would constitute a chain of "unseen universes" grounded in an ultimate or Ideal Ether that would, like the ideal fluid Helmholtz analyzed, effectively be eternal.

Tait's ethereal hierarchy was an attempt to explain scientifically how actual atoms arise, but science wasn't his only concern. All these unseen universes also served a theological purpose, for they suggested that human beings too might be grounded in an Ultimate Reality. As a good Scottish Protestant, Tait speculated that the Ideal Ether was the realm of the Christian soul. Between the world of mundane matter and this soul-filled Ideal were the ever finer levels of ether that Tait now proposed might be a natural home for angels. The Ideal Ether itself would be the natural home of God.

The scientific community was unimpressed by Tait's cascade of worlds, and in a withering critique in the *Fortnightly Review*, the mathematician William Kingdon Clifford commented sarcastically on "the reposeful picture of the universal divan, where these intelligent beings while away the tedium of eternity by blowing smoke rings from sixty-three kinds of mouths." Christians weren't much happier, and many believers regarded *The Unseen Universe* as an unwarranted incursion of science into theological territory. Most physicists today agree with Clifford and dismiss vortex atoms as an aberration in the history of science; officially, the topic has been relegated to the dustbin. Yet we should not be so fast to judge, for we shall meet this idea again, and not just in Jim Carter's front yard.

In the meantime, in Jim's yard, he too was thinking about the creation of the universe and how this might be achieved through the machinations of his circlon-shaped particles. Jim didn't believe in God, yet he too recognized that if his circlon-based theory was valid, then it should be able to describe how atoms come to be. Jim had actually worked out his theory of creation in the 1980s in the years after he'd moved back to the Green River Gorge and, like Tait, he had articulated a chain of becoming that was grounded in the existence of idealized particles. Tait's theory had postulated that things begin with an array of idealized vortex rings; in Jim's universe, the cosmic sequence begins with just two massive circlon-shaped rings. Together containing all the matter and energy of our world, these two particles join to form what Jim calls "the primordial atom," a concentrated yet simple structure that ultimately gives birth, through a long sequence of splittings, to all the particles that exist today. Describing this process, Jim makes an analogy with a colony of bacteria "mating and dividing." At each step, the number of particles doubles, with each particle becoming progressively less massive.

Jim had drawn diagrams to illustrate this process in his book, but in 1999 he acquired the tools to bring his theory to life. Just as Maxwell had visibilized the three-smoke-ring encounter with his zoetrope, so, with a high-end laptop and 3-D modeling software, Jim would set about animating his ideas.

"Directing the creation of the universe" was hardly a job to be undertaken lightly, and Jim spent more than a year and a half of sustained effort on the task. First up he had to learn how to use the software to perform such basic tasks as generating solids and moving them through the frame. Throughout the fall of 1999 and into the following year, he worked away at night, setting his Mac to crunch out short bursts of digitally generated video. It was a slow process as he taught himself the techniques—building wire-frame models, choosing lighting setups, manipulating camera angles, and selecting viewpoints. In the beginning the computer would take all night to render a ten-second sequence, so any mistake would necessitate another night of work. In these early tests, Jim experimented with single circlons, which he set spinning against an empty black background, the default setting of the software that also conveniently represented the absolute void of space that Jim's theory dictates as the background to the All.

In these first animation trials, circlons would simply spin in the middle of the frame, but as Jim developed his repertoire of "camera moves," he learned how to make them glide and whoosh through the field of view. As his skill level developed, more and more circlons were added, and soon whole cotillions were dancing through the void. In his books, circlons are usually colored red or blue, but for the animations Jim chose electric green. With their hard, glossy bright finish, these early efforts look like nothing so much as plastic rings from an infant's toy set, and watching these frenetic little films is like seeing a universe envisioned by a hyperactive child.

Jim began to realize that if he wanted to see more of the circlon dynamics, it would help to make the rings transparent. So he tweaked the animation parameters to create a semitranslucent finish, to which he added a software "glow." Now they actually looked like shimmering green smoke rings—translucent and almost fluffy in appearance. Within his computer, Jim had created a "world" in which he could explore the dynamics of circlon science. What would happen, he asked, if he set these virtual circlons to interact with one another?

The first thing he decided to animate was the periodic table. On Catalina he had built his sequence of plywood models to illustrate each element. Now he set himself the task again, building up the pattern of the elements with his animated rings: one ring for hydrogen, two rings for helium, three for beryllium, and so on. In the animation, each ring flies in from the side of the screen and attaches itself to the existing structure that builds up in the center of the frame like a glowing piece of crystal. As his software proficiency improved, Jim began to do "fly-throughs" in and around these structures, pushing the digital point of view into fancy pans and spins. Next he added the protons and neutrons, which appear in the animations as ruby red circles and golden balls vibrating between the emerald rings.

By visibilizing his theory, Jim was making his universe ever more real to himself and to anyone else who cared to see. Not that there were any takers. Although Jim continued to update his book at a feverish pace throughout 1999 and 2000, sometimes printing off new editions every week, no one in the physics mainstream showed the slightest interest. This two-year stretch was one of the most productive periods in Jim's life, and during this time he was consumed with his physics at both practical and theoretical levels. Each time I saw him during these years, he would happily show me his latest "movie," and many were the times he'd set himself "an animation problem" while he went off to sort out a bunged-up drain or to fix some bit of machinery.

With the periodic table under his belt, Jim now applied himself to the challenge of bringing an entire universe into being. In its animated form, Jim's creation story runs about twelve minutes. In the middle there is a long phase in which a cloud of tiny yellow balls drifts through space like interstellar pollen grains. They are meant to be neutrons, and this "neutron era" is the calm before a cosmic-scale cataclysm that is his version of the Big Bang. Only Jim insists it is not a "Bang," but a generalized conflagration "coming from all points at once." His term is the "Grandfire," and in his version of creation, the origin of our universe is not a point, but—*guess what?*—a circlon-shaped cloud. The shattering of this cloud is what marks the birth of matter as we know it, the coming into being of present-day stuff.

After the Grandfire, Jim's theory proposes a universe that evolves in much the same way mainstream physics describes: Neutrons and protons coalesce to form atoms, atoms congregate to form stars, stars condense into galaxies, and so forth. In Jim's story, it's what happens before the Grandfire that is revolutionary, for this seminal event occurs in the sixth phase of his creation process, meaning that there are five epochs before it. The details of what happens in each of these eras are complex, though in the brief account of his animation you can follow along as he narrates the plot.

When the animation begins we see two rings spinning in space, one red and one blue. After hovering apart like a pair of hummingbirds, these two primal particles begin to move toward each other until they join to form a single spinning pair. Jim describes this process as a "mating," between the positively charged (blue) circlon and the negatively charged (red) circlon. As with much of his science, there is an organic bent here. In the animation, this mating is followed by what appears to be a period of gestation, at the end of which the red and blue whirling couple split into *two* pairs. Each of these daughters in turn splits into *two*, which in turn split into *four*, then *eight*, then *sixteen*, and so on. Soon the screen is filled with spinning gametes that whir faster and faster until each red/blue pair spins itself into a yellow ball like a silkworm spinning a cocoon. These are the neutrons that make up the cloud from which the Grandfire will explode.

Though the details of Jim's story are complicated, the overall scheme is clear—it's circlons all the way down. At every stage in the chain of events, circlons issue from circlons that issue from circlons. Although there is only one universe in Jim's scheme, it shares with Tait's scenario a long evolutionary linkage in which some species of idealized rings gradually unfold over time into ever more mundane versions of themselves. Both Carter and Tait ground their cosmology in a belief that the seeds of the present are inherent in the past and that what exists today must be some more mundane version of what has existed before. Tait called this Kantian presupposition "the Principle of Continuity," and in *The Unseen Universe* he insisted that such a continuum was a requirement for advancement in science. "We make scientific progress in the knowledge of things," he wrote, by looking "for its antecedent in some previous state." For Tait indeed, the very possibility of science was premised on our belief that we can proceed along a "path from the known to the unknown, or to speak more precisely, our conviction that there is a path at all."

The idea that our universe has a developmental path that might be traced by rational thought is one of the hallmarks of the modern age. Prior to the eighteenth century, Christians held that our universe was static; natural forms were given by God at the start of time and were categorically unchanging. Kant, from whom Tait took his Principle of Continuity, was one of the first thinkers who attempted to construct an evolutionary account of the material world, an effort that led him to propose the existence of what we now call "galaxies," or, to use his more lovely term, "island universes." Following in Kant's footsteps, Tait tried to understand how *matter* might evolve. Both he and Kant were influenced by an emerging current of thought that saw it as a mission of science to understand not only the forms of our world, but their ontology. In this respect both Kant and Tait were well ahead of their time.

Perhaps it is not a coincidence that Tait's worldview, which grounded reality in an eternal Ideal, would be rejected by a scientific community that was just then in the business of rejecting eternal ideals and embracing instead the concept of continual change. Darwin had shown how life was continually changing, and geologists had shown how the crust of the earth itself was in flux. Yet for most physicists of the nineteenth century, the question of how matter begins wasn't even on the table. It wasn't until well into the twentieth century that such a problem would be embraced by mainstream physical science.

Darwin had used the phrase *descent with modification* to explain the process by which one animal form could change over time into another, and in Jim's universe also we see that simple forms morph into more complex ones over time. Throughout Jim's cosmos we can trace the influence of Darwinian themes, and one of the most striking features of his theory is its deep-seated organicism. Always with his circlons there is a tendency toward pair-bonding—positive with negative, "male" with "female," the "mating" of opposites. Darwin aimed to show how present-day creatures could come into being from prior ones. What he could not show, and did not attempt to, was how life began. The theory of evolution begins with life as a given, and a century and a half later scientists are still speculating about how the first organisms formed. I stress this point because it is possible to look at Jim's theory and think that he has magicked his first circlons into being: If circlons come from circlons that have come from circlons, then where do the *original* circlons come from? Insiders tend to be scathing of outsider moves like this, but this is also one of the problems biblical literalists point to regarding Darwinian evolution—if the ideas cannot explain the origin of life, they say, then should we accept them at all? Ontology isn't just another scientific detail, it is a major philosophical dilemma. Just as Darwin could not account for how life began, so Jim cannot tell us how the first circlons appeared. If that is a fault of his theory, it is one it shares with every other theory of creation, including those given by mainstream physics today.

Contrary to popular myth, the Big Bang does not bring a universe into being out of nothing. In the beginning—*before* the Big Bang (whatever "before" means in the absence of time)—present-day physics posits the existence of a flux called the "quantum vacuum." This seething sea of "nothingness" is filled with "virtual particles" that pop in and out of being so quickly, they are declared by the laws of quantum theory not to exist. The particles are called "virtual" because their lives are so brief, they are considered not to *really* exist. They are being and not being at once. In *this* genesis story, our universe comes into being when one of these "nonexistent" fluctuations acquires so much energy, it rips itself out of the "nothingness" and explodes into existence as a proto-universe. One of the goals of a "Theory of Everything" is to explain how this undifferentiated seed transformed itself into the particles and forces that fill our world today.

More so even than the origin of life, the origin of the universe presents a unique metaphysical challenge. Logically speaking, you cannot get a world going from absolutely nothing. As the ancient Greeks understood, you have to start with something, some substance or force or entity that just simply *is*. Whether we call it the Prime Mover, the Primum Mobile, the First Cause, the Singularity, the Big Bang, the Quantum Vacuum, the Ideal Ether, or the First Circlon, there has to be something that gets the system going. Theories of origination are expressions of faith grounded in beliefs about what constitutes a reasonable premise for a starting point. In all the variations of contemporary physics, theories of the universe begin with the total energy already inherent in the system. In quantum theory, it arises from the foaming of the "vacuum"; in general relativity, it is present in the "fabric" of "spacetime"; in string theory, as we shall see, the preexisting "void" is so fecund, it spontaneously produces not one universe, but a potentially infinite array of universes. For

Peter Guthrie Tait, the original energy was present in an Ideal Ether, while for Jim Carter it is to be found in two original circlon-shaped particles. In each case we are asked to take as a priori an unexplained—and unexplainable—*something* that forms the grounding for all to come.

Chapter Nine

GRAVITY AND LEVITY

"WARNING! YOU ARE not going to like the idea presented here."

So begins an early version of Jim's book. The reference here is not to circlons, but to his new theory of gravity, for it isn't merely that Jim wants to describe gravity differently, he believes Newton's force is a fiction that insiders have collectively made up. Although this is the oldest facet of Jim's theory, it remains by far the most bizarre of his ideas, the one "impossible thing"—as he likes to say—that would clarify so much else, if only we could bring our minds to accept it.

The first time Jim explained gravity to me we were standing on his back porch overlooking the gorge. In the background as we talked, a carload of tourists stopped on the bridge to admire the view and drop sticks in the river below. The waterfall was in full flight, and so my field of view was filled with what seemed irrevocable proof of gravity's power. Jim had his own stick in hand, and one of the dogs could barely contain its excitement, thinking that a game of catch was in the cards. Jim had scheduled his morning so we weren't to be interrupted—there were no cars to be fixed or pipes to be laid or tenants to be dealt with—and in theory he had the whole day to teach me about gravity. Given the calls on Jim's time, his undivided attention is a valuable commodity, and I'd been looking forward to this moment with some anticipation. Although I had read his book, its account of gravity was frankly so weird that I couldn't quite believe it as written. Surely in person he would cast a less alarming light.

Stick in hand, with faded jeans and a checked flannel shirt, Jim presented an unlikely picture of a theoretical physicist. Then again, hadn't Newton himself begun under equally mundane circumstances? Though the story of the apple bopping the great physicist on the head is apocryphal, Newton had been inspired to think about the forces holding the cosmos together by the simple act of falling, an occurrence so much a part of our lives that it is all too easy to pass by as nothing worthy of reflection. Apples, waterfalls, cups of coffee, buttered pieces of toast—they all fall down. What causes this to happen? Newton asked. Why, for instance, doesn't the apple hover in the air? Or float upward? By subjecting the question of falling to the methodologies of science, Newton had brought a new cosmology into being. For Jim Carter, gravity occasions deep questions still. For him the issue is not merely what *causes* an object to fall to the ground, but what actually *happens* when an object "falls."

Standing on his back porch overlooking the gorge, Jim put the question to me with innocent aplomb. "What happens," he asked, "if I let the stick go?"

As he raised the question, he raised his hand and let the stick fall. At least, that's how most physicists would describe the event: "The stick fell to the ground." Not so, said Jim. According to him, it is not the stick that falls, but the earth. "Matter doesn't fall down," he declared, "it's the earth that falls up." According to Jim's theory, if I "drop" a stick, *it* doesn't *go* anywhere; it sits still in space while the surface of the earth rises to meet it. It is "the earth that hits the stick," as it were.

Watching the water tumbling down the canyon of the Green River Gorge, I tried to digest the magnitude of this proposition. The granite sides of the chasm seemed to stand as a testimony to the steadiness of the earth, and I strove hard to imagine all that rock *moving up* toward the stick. In my mind's eye, I tried to visualize the stick sitting still in space while millions of tons of stone rushed up to meet it.

I pushed out of my mind myriad questions that my university science education occasioned me to imagine. If the surface of the earth was constantly moving outward, I thought, what would happen to the moon? What about all the other bodies in our solar system—the sun and the planets, and *their* moons—how come they don't appear to be changing size? Leaving this aside for the moment, I went with the flow of Jim's thoughts. "If the earth is moving up toward the stick," I said, "then something must be propelling it up and out. What power could this be?"

"That's easy," he replied, "our planet is expanding." Every nineteen minutes, he went on, the earth doubles in size, so that, as in the case of an expanding balloon, its surface is constantly pushing out and up. According to Jim, "gravity" is not an invisible "force," as Newton had declared, but the consequence of a continually expanding earth. Our feet remain stuck to the ground because the earth's surface is pushing up against them. In short, "gravity" is not a downward "force," but an upwardly acting "pressure."

At this point in his narrative, the dogs demanded that the

stick move toward the front lawn, and Jim hurled the instrument of his physics lesson over their heads and into the trees, leaving my head reeling with the consequences of this bizarre idea. Above the spectacle of the waterfall, I forced myself to imagine our planet ballooning out, the rock and trees and water, too, pouring themselves into the surrounding space. If, as Jim claimed, the size of the earth was doubling every nineteen minutes, then in the time he'd been explaining gravity to me, our planet had grown half as large again as when we'd begun. What hidden feature of our universe could be responsible for this extraordinary inflation? Naturally Jim had an answer here as well: It was an inherent quality of matter. In Jim's theory, *all matter* is expanding; every atom is constantly increasing in size, so that everything in the universe is getting bigger and Bigger and BIGGER. Every particle, every planet, every star, is in a perpetual state of expansion. It would have to be, because if the earth were the only thing doubling in size, it would quickly overwhelm the rest of the cosmos. The only way Jim's idea makes sense is if *everything* is getting bigger. You, me, the chair you are sitting on, the clouds in the sky—according to Jim's theory, we are all continually doubling in size.

By far the most unorthodox aspect of his science, Jim's theory of gravity seems to surprise even him. Thirty years after he developed his ideas on "expanding matter," he still prefaces every conversation on the subject with a qualification to prepare his listener for what he acknowledges is a nearly abhorrent idea. In his book the **WARNING!** is followed by a statement of exactly how appalled he expects his reader to be. "Your reaction of immediate rejection will occur at the moment you first realize what this idea is," he confides. "Your feelings will be more like moral indignation than passive disbelief in a logically flawed line of thought." But immediately after this olive branch comes a challenge to critics to prove the idea wrong. It is one of the most audacious passages Jim has written and among the more confrontational. It is the work of a very young man at the peak of his creative powers who has not yet been tempered by decades of academic indifference. It is a passage at once charming, cheeky, and arrogant, and however else it might be construed, it is a battle cry from the fringe.

> If you, as an intelligent person believing in free will, feel insulted that I would so arrogantly prejudge your opinion on such an important and controversial issue then I challenge you to suppress your emotions at least long enough to allow your intellect to establish at least one well reasoned argument, based on experimental results, as to why this idea could not be true. If you can create such an argument, before being overcome by what I feel is a genetic predisposition to push this idea out of your consciousness, I would love to see it, because you will have accomplished something that no one else has been able to do.

Jim's cry against gravity is not confined to his science, for much of his life may be read as a series of triumphs over this supposedly inevitable force. Where digging serves as one theme in an effort to understand his life, gravity suggests another organizing principle. Over the past thirty years, overcoming gravity has indeed served as the foundation of his income. Though many people are surprised to hear so, Jim makes a pretty good living, *so* good that he and Linda can afford to take holidays almost any time they choose, and in the years I have known them they have traveled to Russia, Greece, Italy, England, Mexico, China, Tahiti, the Bahamas, and Peru. In conversations with a wide range of people, I have often found that many imagine "crank" theorists to be poor, as if outsider status academically ipso facto implies the economic equivalent. I confess this was my own presupposition. Yet outsiders inhabit a huge range of economic brackets, and wealth is no barometer of intellectual leaning.

By far the most famous outsider physicist today is a wealthy man. Stephen Wolfram is his name, and aside from being one of the youngest people to win a MacArthur Prize Fellowship, he was a mathematical prodigy who graduated with a Ph.D. from Caltech at the age of twenty. After he won the MacArthur, Wolfram left academia to forge a path of his own and made a fortune developing a software package used by scientists all over the world. Wolfram's wealth enabled him to pursue work on his own "Theory of Everything" unencumbered by the demands of academic tenure, and in 2002, he presented his ideas in a twelve-hundred-page, self-published book called *A New Kind of Science* in which he offered more even than Jim Carter has dared to claim. Encompassed in Wolfram's theory was a new explanation for space and time, a new explanation for the laws of thermodynamics, a new account of the creation of the universe, an explanation for the origin life, and his own account of free will. I have never seen a more comprehensive theory or one that has incited so much irritation in the academic class. Unlike Jim's book, Wolfram's work could not be entirely ignored by professional physicists, for the media loved the story of a child-genius-turned-renegade and happily devoted hundreds of articles to his book.[1]

Jim Carter hasn't made a fortune, but like Stephen Wolfram, he has built a successful company around his own invention. The "lift bag" business, as he calls it, operates out of the old emerald lodge near the bridge and employs half a dozen local men and women from the Enumclaw–Black Diamond area.

The very name "lift bags" connotes optimism, and the invention arises from what turns out to be a widely shared desire to overcome gravity. Anyone who owns a boat that has sunk, anyone who has dropped something on the ocean floor and wants to retrieve it, anyone who just likes picking stuff up in the sea, has a use for Jim's invention, and the Carter Lift Bag Company is a thriving business.

Jim dreamed up the idea of lift bags when he was working on Catalina. If *you* had to haul two-hundred-pound bags of abalone through the water by hand, you too might be inspired to think about options. In principle, the concept couldn't be simpler, and the science underlying it had been known for more than two thousand years. Why does an object float in water? Archimedes asked. Because it displaces its own weight with a volume of water that would weigh the same. Legend has it that Archimedes came to this insight in the bath, and fortunes have been made selling Archimedean bath toys. How many toddlers have experienced the joy of drowning a plastic duck over and over again? If you take one under the water, it bobs back up to the surface as soon as it's released. Jim realized that this innocent game held the answer to his need. Let us say one wants to lift a two-hundred-pound net of abalone. Jim saw that, in effect, all he had to do was tie it to a plastic duck. As a diver with a scuba tank, he realized he could take an inflatable bag under the water, attach it to the net, inflate the bag, and presto, the abalone would be carried to the surface. Every cubic foot of air displaces a cubic foot of seawater, which weighs sixty-three pounds, so for a two-hundred-pound load, the bag would need a volume of a little more than three cubic feet. For a two-thousand-pound load, the bag would need to have a volume of thirty-two cubic feet. In a liquid environment, gravity can literally be overcome with a puff of air.

Given that under marine law boat owners are required to raise any of their vessels that sink, the lift bag business has a huge built-in clientele, and every time there is a hurricane in the Caribbean, the little factory at the gorge can't turn them out fast enough. Depending on your needs, you can buy a lift bag to raise anything from a huge sunken ship to an old anchor.

Jim didn't set out to profit from other people's misfortune. When he invented lift bags, he was simply trying to make things easier for himself. With two young sons to raise and a theory of the universe to develop, he had better ways to spend his time than lugging fish. So he hunkered down and did the research. It wasn't rocket science, but it did require a good deal of experimentation. He had recognized that in principle a scuba tank could be used for a purpose other than the one for which it was designed. How could he get this to work? The evolutionary biologist Stephen Jay Gould coined the word *exaptation* to describe this process of sideways functionality in the organic world and had noted how often human technology also develops this way. With his own exaptation, Jim faced a slew of engineering challenges: What kinds of valves would best connect to the scuba hose? How could the bags be sealed properly? What materials should he use? What designs were best for underwater maneuverability?

The first lift bag he made was nothing more than an old duffel bag sealed with glue, and as with all good ideas, inspiration had to be carried through with a good deal of perspiration and problem solving. Linda became part of the project, too, and it was she who worked the sewing machine stitching up the samples. In the little house on Catalina, the pair of them sat around their kitchen table and tried things out, and over a couple of years, as Jim recounts the story, they "learned how to make a pretty good bag."

Jim had invented lift bags for his personal use, but soon his abalone-diving friends were wanting them, too. Then the dive store on Catalina decided to stock them. Recreational divers used the bags for lifting treasures such as old anchors and giant clam shells. When Jim and Linda went on vacations, they started taking along a carload of bags to sell to dive stores along the way. After a few years, it turned out that Jim was making more money selling lift bags than he was from selling abalone, which was fortunate because the abalone had begun to run out. By the end of the 1970s, the great banks that once stretched the length of California had been fished out, and eventually commercial abalone diving was banned on Catalina. The rise of the lift bag business and the decline of the abalone trade precipitated Jim's decision to return to Washington State. Now that California was no longer essential to his income, he was free to move anywhere, and he chose to return to Enumclaw to establish his little factory. Ever since, the Carter Lift Bag Company has provided a steady income for both his family and his partners, a local couple named Dave and Diana Blackburn.

When you take things under the water, it's a good idea if they are brightly colored so they don't get lost in the dim light, and

Carter lift bags are produced in a variety of vivid pop art hues—bright red, canary yellow, brilliant orange, and lime green—all nattily embellished with black plastic fittings. There are long thin snakelike tubes used as "Personal Floats" to mark a diver's position in the water, squat cushiony "Pillow Bags" for small-scale salvage operations, and giant-scale "Salvage Sausages" that can lift up to ten thousand pounds. Now a standard Carter company item, Salvage Sausages were originally designed as a special order for the U.S. Navy "to recover parts that had dropped off the back of rockets." The U.S. Navy, the Coast Guard, and boat owners the world over use Carter bags. After the Gulf War, Middle Eastern millionaires ordered dozens of Jim's bags to raise the yachts that had sunk during the fighting. Given the popularity of nautical sports, the lift bag business has far exceeded Jim's expectations, and if he would allow it to happen, the company could be many times larger than it is now. Jim has resisted going down that path, for as with so many aspects of his life, he prefers to operate on a small scale with a group of people he trusts. The modest size of the business also guarantees that he has plenty of time for his science. If one wants to be reconfiguring the foundations of physics one can't be supervising production orders sixteen hours a day.

One way of looking at the cosmos Jim has constructed is as an extension of the world he has built for himself at the gorge. In both cases, his "universe" is governed by a calculus that places an intimate set of relationships at its core: on the home front, his wife, his children, and the friends who share in his company; on the physics front, the circlon-shaped particles, whose groupings so resemble a family dynamics of their own. With lift bags, as with circlons, small seeds of potential inflate into a self-sustaining whole. Here too the gravity-defying invention of his bags mirrors the gravity-denying thesis of his science.

In the scheme of Jim's science, circlons offer an alternative to the quantum mechanical theory of matter and they also provide a substitute for special relativity, which is Einstein's theory of motion. "Expanding matter" is Jim's alternative to general relativity, which is Einstein's theory of gravity. General relativity is one of the more powerful theories in science. Here, also, gravity is seen not as an independent "force," as Newton proposed, but rather as a quality built-in to the very fabric of reality. Einstein's equations explain gravity as a property of spacetime itself, which is often described metaphorically as a vast rubber membrane that bends and flexes in response to large clumps of matter such as the earth. According to general relativity, what we experience as the downward acting force of gravity is actually the stretching and flexing of this spacetime sheet. A large body like our planet acts on the membrane of spacetime rather like a bowling ball sitting on a trampoline—in both cases, the massive object creates a depression in the surface, warping the weft and weave of the "fabric" around itself. In the case of our universe, the warp and weft are literally space and time.

General relativity is generally regarded by insiders as one of the most elegant theories in the history of science, yet it inspires among outsiders a particular sense of ire. Fringe physicists tend to abhor Einstein's masterwork, however, it has not been the subject of nearly so many of their refutations as special relativity has provoked. While half the works in my collection offer alternatives to special relativity, it is much harder to come up with alternative explanations for gravity, a fact that insiders acknowledge as well. Among academic physicists, gravity is widely regarded as one of the hardest of all problems. While insiders don't doubt that general relativity will form some part of any Ultimate Theory, they too believe there are concepts underpinning gravity that we don't yet understand. Finding a theory of gravity that is compatible with quantum ideas is now regarded as one of the most pressing problems in theoretical physics.

Outsiders tend to dislike general relativity because more so even than quantum theory it is nearly impossible to understand without knowing the math. The equations of general relativity are ten-dimensional "tensor" equations whose abstract nature offers formidable challenges even to highly trained professionals. Many physicists don't understand much about the subject, either, and despite all the popular books purporting to explain its ideas to the average citizen, there really is no way to comprehend the relativistic view of gravity if you don't understand a fair bit of mathematics. Of all the outsiders I have met, Jim is one of the few who refuse to be intimated by gravity and who actually have a concrete idea about how it might work. With relativity, Einstein refuted the notion of *absolute* space and time and insisted that these qualities change in relation to each other; Jim insists that space is absolute but that matter and time both "change proportions." "Something changes," he says. "Is it matter or is it space?" According to Jim, mainstream physicists have gotten it backward. He talks about his own idea as a "mirror image" to general relativity and sees himself as a kind of "anti-Einstein."

In the 1930s and 1940s, the philosopher Karl Popper refined the idea of what it means to demonstrate scientific truth. Popper proposed that a "scientific" theory is one that can be experimentally put to the test and potentially proven false. According to

Popper, if one wants to claim an idea belongs to the realm of "science," then it has to be possible to do an experiment that might prove it wrong. Of course, every scientist hopes his or her idea will be proved right, but as Popper saw it, the possibility of demonstrating falseness is what makes science different from other domains of knowledge such as religion or mythology. By Popper's definition, Jim's idea is a genuine scientific theory because he has proposed an experiment that could refute it. With a large ball bearing and a heavy metal tube taken up on the space shuttle, we could in fact determine whether his theory is wrong.

During the years I have known Jim, I have sometimes wondered what would happen if NASA provided a service for testing outsider ideas. Jim once told me he thought there "should be an office where you could send in a theory and have it evaluated by experts." Outsiders like him could get some feedback. At the moment, he said, "everyone just says that Einstein is right and if you say Einstein is wrong, no one will listen." It is a refrain I have heard from many other outsiders, along with a desire for feedback.

NASA has in fact always received a lot of way-out proposals, but like any government agency, it doesn't have the facilities to take on every idea. No university physics department does, either. (One famous physicist told me he receives so many outsider theories that if responded to them all, he would have no time to engage in his own research.) All institutions must pick and choose the projects they take on, and the purpose of peer-review panels is to set standards and guidelines for what a given institution will do. Peer-review panels can often be rather cautious, especially when multibillion-dollar facilities are at stake, and one of the gripes many *insiders* have is that bold ideas from the mainstream *also* get screened out. So it was a rather brave move when in 1996 NASA set up a program whose stated goal *was* to support way-out ideas. Called the Breakthrough Propulsion Physics Project (BPPP), the program aimed to test wild ideas from the fringes of physics that might help advance the next generation of space-ships. NASA was looking for fringe ideas for the reason that if we humans are ever going to travel to the stars, we are going to need a radically new kind of propulsion system; rocket fuel simply can't deliver enough energy for interstellar transport. If we ever hope to realize the *Star Trek* fantasy of going to the stars, we do actually need something like a warp drive.

In the six years of its "active funding," the BPPP supported eight topics of research, several of them variants on the idea that we might be able to suck energy out of the quantum vacuum and gain access to an infinite source of energy. The idea is one that the majority of mainstream physicists regard as delusional, yet officials at the BPPP decided that their mandate required them to at least be open to the possibility. No one has actually proven this is impossible, and there have been many instances in the history of science when "impossible" things have turned out to be real. Atomic bombs and orbiting satellites were both once deemed beyond the realm of possibility. Researchers supported by the BPPP worked hard to make their cases plausible. Papers were published in respectable, if rather obscure, journals. My favorite was one published in the *Foundations of Physics* entitled "A Gedunken Spacecraft That Operates Using the Quantum Vacuum (Adiabatic Casimir Effect)." In plain English, that translates as an "imaginary spaceship that operates on the energy of empty space." For those who might have trouble envisioning such a vehicle, the BPPP pages on NASA's Web site provide a helpful illustration of a torpedo-shaped spacecraft zooming into a wormhole.

In all, NASA devoted $1.6 million to the BPPP project, paltry by the standards of particle accelerators and deep-space telescopes, but nonetheless a sizable slice of taxpayer largesse. In 2003, after heavy criticism from the mainstream physics community, NASA stopped "active funding" but continued to support the program's office until 2008. The agency's final contribution was to assist in the compilation of a graduate-level technical book called *Frontiers of Propulsion Science.* Sadly, the Web site reports, "No breakthroughs appear imminent." "Objectively, the desired breakthroughs might turn out to be impossible," the text continues, "but progress is not made by conceding defeat."

Progress is not made by conceding defeat.

It bears repeating. Here we have the outsider's creed exactly. The pedantic pessimists, the negative-minded literalists, those who would be intimidated by the laws of nature, they stand in the way of humanity's future. Whatever you can say about outsiders, they are optimists to the core. Managing that optimism was program director Marc Millis's primary challenge. Millis's problem as head of the BPPP wasn't the lack of proposals but rather the tsunami of Great Ideas that came flooding through his door. Proposals arrived from all over the planet and from across the academic spectrum: from tenured professors of physics, credentialed engineers, aerospace professionals, antigravity theorists, and backyard hobbyists. Millis's most formidable task was trying to develop some criteria for assessment. "On a topic this visionary and whose implications are profound, there is a risk of encountering premature conclusions in the literature," he wrote tactfully on the Web site. Yet premature conclusions were precisely what his colleagues found so appalling and why, after several years of trying to stave off ridicule, NASA withdrew its support.

As a science journalist in the 1990s, I followed Millis's endeavor from afar and secretly hoped NASA would channel a lot more funds his way. I was rather excited to know that my tax dollars were working in so subversive a fashion. At one point when the guillotine was beginning to fall in 2003, I called Millis up and we had a heartrending talk. He was still hoping to save his ship and was enthusiastic about the possibility of me writing about the project in my book. The problem *I* had with the Breakthrough Propulsion Physics Project was deciding whether or not it counted as an outsider endeavor. Millis was making every effort to make it seem as if it weren't, for he knew that continued funding depended on the appearance of respectability. On the other hand, his stated mission was to boldly go where no physicist had gone before. He was caught between a warp drive and a wormhole, and frankly he couldn't really win.

Jim Carter was not among those who sent the BPPP a proposal, because Jim is one of the rare outsiders I've encountered who insist that space travel isn't possible. A whole section of his book is devoted to explaining why. From a star tripper's perspective, Jim's theory of gravity is a downer. Jim may not have written to Millis, but he does have an experiment he wants NASA to perform that he believes would show that general relativity is wrong and that matter *is* expanding. It could be done on the space shuttle with a heavy metal tube and a small metal sphere. The experiment predicts something about how these two objects would behave in relation to each other out in space, away from the gravitational effect of the earth. In practice, it would be expensive because space shuttle cargo costs millions of dollars a pound, which is why so few experiments are done in space and why the available research slots are always subject to intense competition. Jim hasn't put in a proposal to NASA, but he has outlined the experiment in his book. One day, he hopes, someone will put it to the test.

Part Three

SCIENCES OF IMAGINARY SOLUTIONS

Imagination is more important than knowledge.
Knowledge is finite, imagination encircles the world.

—ALBERT EINSTEIN

Chapter Ten

A REFORMATION OF SCIENCE?

JIM CARTER MAY not have submitted a proposal to the Breakthrough Propulsion Physics Project, but the hundreds of people who did are proof that at the dawn of the new millennium, the paradoxical classes are thriving. "Squarers of the circle, trisectors of the angle, duplicators of the cube, constructors of perpetual motion, subverters of gravitation, stagnators of the earth, builders of the universe," they have not gone away. Hundreds of "examples of scientific logic gone haywire" poured into Marc Millis's office, amassing into a De Morgan–esque archive of the contemporary discoverer's mind. I do not know if NASA has attempted to preserve these papers, though I certainly hope someone has seen fit to hold on to this survey of intellectual "debris," for it will surely serve historians of the future as a unique window into the state of physics today.

From a mainstream physicist's perspective, perhaps the most surprising feature of the BPPP papers is that they exist at all. It has been 140 years since De Morgan published his *Budget*, and in that time physics has advanced from its adolescence into the most mature of the sciences, with practitioners talking openly now about a "final theory" potentially being realized in our lifetimes. I think if we had asked physicists of De Morgan's time what they might predict for their science in the public imagination of the twenty-first century, a good percentage would have expected general acceptance. To De Morgan and his peers in the mid–nineteenth century, Peter Guthrie Tait and William Thomson among them, physics was progressing by leaps and bounds, and on almost every front heroic developments were afoot. The public seemed hungry to know more, especially once technological miracles born of theoretical abstractions began to appear—electricity, telegraph, radio, internal combustion engines. Surely these trends would result in widespread assent for the scientific worldview?

We don't actually have a survey of Victorian physicists' views about future public opinion, but we do have something intriguingly suggestive: At the end of the nineteenth century, many champions of science began to predict that during the twentieth century, *religion* would disappear. According to this camp, the advances of science would so overwhelm any competing system of thought that nothing could stand in its way. Tait had written *The Unseen Universe* as a retort to this trend and to propose a way in which physics could be synthesized with Christian beliefs. *The Unseen Universe* failed to capture the imaginations of either scientists or religious believers, and throughout much of the twentieth century science did indeed appear to be driving out other systems of belief. So much so that in the 1980s, physicists like Stephen Hawking and Paul Davies began to speak openly about God being replaced by equations. Yet the death of religion was prematurely declared; if anything, at the start of the twenty-first century it is champions of science who are now on the defensive. Competing systems of thought are flourishing—in religion as well as philosophy, and in the arena of science.

In many ways there has never been a better time to be a fringe physicist. Although insiders remain as unmoved as ever by outsider ideas, paradoxers of a physical stripe have a source of support today that was unimaginable to their predecessors, for they effectively have their own union. Established by "discoverers" themselves, the Natural Philosophy Alliance serves as a publicly open forum in which outsiders can publish their ideas without fear of censure. *Anyone* with a theory of physics can join the NPA and present papers at its conferences. Any alternative theorist can post links to his or her books on the organization's Web site and have his or her ideas detailed on a dedicated Web page. Whatever else they may be, fringe physicists today are demonstrably not alone. Unlike the men who came knocking at De Morgan's door, contemporary paradoxers have a means by which they can comfort and sustain one another.

On the NPA's Web site (www.worldnpa.org), more than nineteen hundred "dissident physicists" are cataloged. This ever-growing roster is listed in what the organization calls its World Science Database, a vast searchable archive containing an encyclopedic wealth of information about theorists and their ideas. Clicking on an entry for any individual brings up a Web page that contains personal contact details, a professional biography, an educational history, an outline of theories, books written, a list of papers presented at NPA events, and, remarkably, an abstract of *all* such papers given. Jim Carter's NPA page lists no fewer than twenty-three conference papers, presented between 1994 and 2010. Here also we find a list of all the NPA events Jim has attended, so we can see at a glance that he has been an active member ever since the alliance's founding year. Not since De Morgan has there been anything like such a catalog, and De Morgan himself would likely have been astounded at the sheer thoroughness with which these data have been compiled. David Scott de Hilster, founder of the World Science Database and the driving force behind the NPA site, has taken it as his mission to establish as near as is functionally possible a complete library of physics outsiders today.

Along with theorists, the World Science Database contains a separate catalog of books these people have written. As of April 2011, more than thirteen hundred are listed, with a dedicated Web page for each publication that includes a description of its contents and links to Web sites where the volume can be purchased. For theorists looking to *publish* a book, the NPA has sourced a cheap, print-on-demand service where would-be authors are assured they will have to pay for only a single copy. Perhaps most impressive, the World Science Database catalogs more than fifty-five hundred dissident physics papers, each of which also has its own Web page.

De Hilster's achievement is all the more impressive because in addition to compiling this vast storehouse of information, he is attempting to bring some order to it. Dissident's ideas are classified into major categorical types, making the World Science Database the first attempt at a Linnaean taxonomy of physics outsiders. Some categories are terms insiders accept, such as Relativity, Quantum Theory, and Cosmology. Others signal that we are beyond academic norms: Ether, Aether, Antigravity,

Pushing Gravity, Zero Point Energy, and Cold Fusion are among this class. Looking at Jim's entry, we find that his theory is classified under Gravity, Nuclear Structure, and Toroidal Ring. Interestingly, his ideas are *not* included in the category of Expanding Earth, a designation that *is* applied to several nearby entries in the database. I was amazed to find that no fewer than seventy-three theories are in the Expanding Earth class. It is indeed one of the NPA's more common themes.

All information in the World Science Database is searchable by author, title, date, and "theory description," enabling anyone who is interested to find out who is saying what about any fringe idea. Wandering through this archive of dissidence, one is struck by the strength of humanity's paradoxical powers. I do not think even De Morgan, gently chiding his discoverers, could have imagined how enduring these forces would be. What would he have made of the fact that toroidal ring theory is now a standardized term in the paradoxer's lexicon? One wonders also what Tait and Thomson might say on seeing their beloved idea diffused into a fog of speculation. Using de Hilster's software, theorists can input their own descriptions of their ideas, so that in place of De Morgan's random perambulations through the fringes of science, here we have a collectively produced encyclopedia that enables dissident physicists to tell us in their own words what exactly they think. The tables have been turned. Instead of an insider's comments on the outside, the World Science Database give us outsiders' analysis of the inside. This is a new *Budget of Paradoxes*, written by discoverers themselves.

The idea of a society of outsiders may at first seem oxymoronic. By definition, "outsiders" stand alone, and the tension is one that NPA literature openly acknowledges. "We agree unanimously on little more than that something is drastically wrong in contemporary physics and cosmology," the Web site explains. For members of the NPA, not even the basic laws of nature are shared, or even initial assumptions about how such laws are derived. *Constructively* speaking, nothing unites these men and women, and at first glance it might seem that little could be more desirable to any of *them* than a mutual society, a forum in which they might all be mutually frustrated: "How can *you* have the true and final theory of the universe when *I* have already found it?" Yet if NPA members do not share a view of how the problems of theoretical physics should be resolved, they are remarkably unified in their view that such problems *exist*. As the Web site's home page states the case: "The Natural Philosophy Alliance is devoted mainly to broad-ranging, fully open-minded criticism, at the most fundamental levels, of the often irrational and unrealistic doctrines of modern physics and cosmology." "The ultimate replacement of these doctrines by much sounder ideas" is the NPA's goal.

The very name of the organization speaks volumes about its aims, for the term "natural philosophy" is a designation that harks back to the era before science became professionalized. As the NPA Web site reminds us, "Newton was a natural philosopher." The word "physicist" was coined in 1840 but by the end of the nineteenth century it had become such a loaded appellation that "amateur physicist" could pretty much be taken as a contradiction in terms. "Theoretical physics" in particular has been viewed ever since as an apex of academic specialization, a discipline that by its very nature supposedly remains the purview of an educated elite. By going back to the older terminology, the NPA seeks to return the field to a more democratic ethos.

NPA members may not agree on how to solve the problems of theoretical physics, but their articulation of its failings is remarkably consistent. Again and again they rail against reliance on mathematics. On the Web site, a prominently placed essay entitled "Our Minimum Consensus" serves as what amounts to a group manifesto. "In science," it begins, "it is all too easy to jump to conclusions . . . [While] math is a wonderful and most valuable help in physics, pure numbers don't give us physics." Only when math is "used properly" is it a legitimate tool in science; otherwise, "catastrophe" ensues, leading to the kinds of "cult theories" that have overtaken the subject in the past hundred years. Author Peter Marquardt does not name specifically relativity or quantum theory in his list of physics' sins, but his writing makes it clear that he considers NPA readers too well versed in this litany to need the enemy named. Just as numerological coincidences enchanted astronomers of the sixteenth century, so Marquardt declares that theoretical physicists today are being blinded by the "pseudosuccesses" of their mathematical tools. The NPA's mission is "to rid physics of this burden" and put it "back on track where it was once a real science" in the mode in which it was practiced before the grip of math took hold.

The NPA has not always appeared as such a beacon of self-confidence. When I first encountered the organization in 1998, the primary tone of its literature was a sort of wounded pride. In those days, the group's only form of communication was a newsletter hand-typed and hand-copied onto pastel-colored papers. To get the newsletter, you had to get onto the mailing list, and it wasn't easy to see how that was done. I had learned about the NPA from Jim, who had been a member since the organization's inception in 1994, before there was even an official name. Jim had helped to work out what the name would be, a process that was taking place in the months after I received his own original mail-out. Though I did not know it then, at the very time I encountered Jim the NPA was forming, and one of the most treasured items in my collection is a selection of its early announcements. Jim has been to nearly every one of the alliance's meetings and has seen the organization transformed into its present Internet glory from what was, at times, a state of near collapse.

The NPA was actually formed by a philosopher, Dr. John Chappell, who found when he was doing his doctoral dissertation in the philosophy of science that his advisers could not or would not deal with his criticisms of Einstein. As a dissident, Chappell had tried to go it alone, but the experience was so isolating it provoked him into starting the NPA as a means to gain some moral support. By 1998 when I met him, he had managed to pull together a small band of dissenters with whom he kept in touch through irregular mail-outs and via an annual meeting where members could present their ideas. Given the resources available to Chappell, those early events were little short of a miracle. He was so hard up, he couldn't afford airfares to his own events and had to rely on the Greyhound bus when he commuted from his home in San Luis Obispo. Chappell extended his pecuniary sympathy to his members: NPA dues were set at a mere $20, and even that was waived "for those in extreme hardship or in nations with an extreme scarcity of U.S. dollars."

At the time I met Chappell in May 1998, the future of the NPA seemed uncertain. For the past three years, he had been trying to get the group accepted into the annual meeting of the American Association for the Advancement of Science (AAAS), the largest and most prestigious event on the American science calendar. Each year he had been sending in proposals for NPA papers to be presented as a formal conference "track," and each time he had been turned down. No reasons were given and none were required; in the process of peer review, the NPA's proposals had simply been deemed unworthy. That year proved to be no exception, and at the AAAS meeting in February, the NPA had again been absent from the program.

In the March 1998 edition of his newsletter, Chappell reported to his members on this sad state of affairs. In the face of rejection, the NPA had decided to organize its own event, which it staged at a local high school five blocks away from the vast AAAS convention in downtown Philadelphia. Eleven dissident papers were presented. At the AAAS event itself, NPA members distributed flyers and were on hand to explain about the alternative events happening down the road, but as Chappell noted, "All this attracted very few listeners, and no press coverage." Chappell was a patient man and not one to give up easily. "I sensed more curiosity than disdain among those who saw our flier," his newsletter reports. Nonetheless, he was becoming a little discouraged. He had first approached the AAAS committee as a lone dissident in 1979; by 1998 he was sixty-five and in frail health, and his spirit was clearly flagging. His report ended on a plaintive note. "Should we keep trying?" he asked.

In May that year, I attended the NPA's annual conference with Jim. Although the AAAS national committee had refused to allow the dissidents to nestle under its umbrella, Chappell's proposal had been accepted by a regional AAAS division, the Southwestern and Rocky Mountain division, or SWARM, as it is known. That year, SWARM was holding its regional conference in Grand Junction, Colorado, and its leadership graciously allowed the NPA to tag along. As it turned out, a senior scientist on the SWARM committee had a dissident in his family and did not see any harm in giving outsiders a room to hash out their theories on their own.

In Grand Junction, the sense of hurt sustained in Philadelphia lingered over the proceedings, but like many wounded creatures, Chappell was keeping up an air of defiance. He himself delivered the opening address, entitled "Coping with Suppression of Dissident Thought," which began with an indictment of "irrationality" in physics and concluded with an almost painful discussion of how outsiders could cope psychologically with the lack of feedback from the mainstream. Yet even Chappell acknowledged that dissidents often had a ways to go with presentation standards, and one of his aims in Grand Junction was to move his members toward a more "professional" level of discourse. In particular Chappell wanted to get away from the all too common tendency of members to claim that they had resolved all major problems in physics. In the "Call for Papers" leading up to the event, he had stressed that presentations must be "open-ended" and "leave room for discussion." Speakers were required to submit papers "concentrating primarily on a problem that the author has *not* fully resolved." It was all part of an effort, as he put it, "to get out of the rut of authors just giving their own ideas and paying little attention" to anyone else. In case that wasn't clear enough, Chappell went on to note that the aim of the meeting was to "discourage excessively prideful claims that 'I have the best and final answers,' which usually turns out not to be true, and usually is best left to others to recognize."

And herein lies the central tension for the NPA: At the Grand Junction meeting, I sat through almost thirty talks, each claiming to present a key to Ultimate Reality. Given Chappell's directives, most speakers were trying hard to leave room for doubt, but you could tell nobody had his or her heart in that. Among the presenters was Dr. Domina Eberle Spencer, one of the alliances few female members, who would later go on to serve as its president and whose ideas on electrodynamics firmly refute Einstein. It was pretty obvious in the question time following each presentation that people were struggling to keep up a discussion on one another's work. Everyone was itching for his or her own turn at the podium. Thinking about this in my hotel room after two days of talks, I recalled the book *The Three Christs of Ypsilanti*, an account of an experiment conducted in the 1960s at the Ypsilanti State Hospital in Michigan in which three schizophrenic patients, all of whom claimed to be Jesus Christ, were put together in a room. In Grand Junction, I realized I had been watching thirty Jesus Christs. Everybody had the Answer. Everybody was the One. To their credit, the NPA theorists remained polite to one another and nobody openly challenged anyone else, but you could feel the disconnect among them and the general air of bewilderment: How exactly is a person supposed to respond to someone *else's* harebrained theory when each person has his or her own Solution?

That evening, with a spectacular sunset as our backdrop, Jim and I went for a walk to discuss the day's events before joining the other speakers at a local Chinese restaurant. It was the first time I had seen him in the company of other theorists, and I wasn't sure how he might be taking all this. As it happened, Jim was as amused as I was. Unlike many of the other speakers, he had delivered his paper coolly, with very little emotion. He didn't seem to be trying to convince anyone of anything. "This is it and here it is," was his cut-and-dried approach. It was equally obvious that even among this crowd he was outside the fold. Most of the other speakers were ether theorists, so in the question time following Jim's talk nobody had much to say. How could they, when their own theories differed so wildly from his? Jim wasn't perturbed. This was his fifth NPA conference, and he knew what he'd be facing. Puzzlement and incomprehension: These were the norms he had come to expect from the Natural Philosophy Alliance.

In 2002, John Chappell passed away. He was sixty-eight years old and looked a decade older. Although he did not seem bitter, he was enormously frustrated and he never comprehended how the world could be so indifferent to his ideas. He had spent his intellectual life trying to open up channels that, to quote Michel Foucault, seemed "too narrow, skimpy, quasi-monopolistic, and insufficient." He had dreamed of "a new age of curiosity" and had seen firsthand that "the things to be known are infinite." As the NPA membership attested, "the people who [could] employ themselves at this task" existed. Chappell did not live to see the fruits of his vision realized, but the World Science Database is the legacy of what he began.

In 2010, I again attended the NPA's annual conference with Jim. This time, the event was held at California State University in Long Beach; 121 papers were presented from dissident researchers as far away as Russia, Portugal, India, and Australia. The accompanying book of papers ran to more than seven hundred pages. Speakers presented their talks using PowerPoint and other software programs. Video projectors beamed presentations onto giant screens, and many speakers showed animations to visualize their ideas. A camera crew was on hand to videotape the events, and some of the talks are now available on the NPA Web site. An exhibition area was set aside so that theorists could display models of their ideas and also sell their books. There was a public open day, with talks designed to wow a general audience about the spectacular possibilities all this "new science" would soon make possible. But more impressive than any of these physical manifestations of the NPA's vitality was the way in which members interacted with one another; in contrast with the limp discussions in Grand Junction, the question-and-answer sessions following many of the talks at Long Beach were charged with energy. Speakers questioned one another and cited previous research, they challenged one another on points of data and reminded their audience of pioneering achievements. In short, they acted like scientists at any professional conference.

At the end of the Long Beach event, a group of NPA members gathered on the steps of the campus's giant glass pyramid for a photo opportunity. On hand was a small coterie of early NPA members, including Jim and Steven Rado, the Hungarian ether theorist whom I had met in 1999. On the section of the NPA Web site devoted to the Long Beach conference is a message from Rado expressing his delight at the proceedings and his pride in being a member of the NPA. Rado's message has lapses in grammar, reflecting his origins as a non-native speaker of English, but its passion and enthusiasm and sense of triumph shine radiantly through. I can think of no better testimony to the resurgent power of the paradoxical spirit than to quote it in full:

> With my wife, Rochelle, we spent three unforgettably gorgeous days in the beautiful Long Beach State College [*sic*] and a very busy public day in the fantastic Pyramid Pointe. What can we say that would properly express our overwhelming feelings about this adventure? What would my first good dissident friend, the founder of the Natural Philosophy Alliance, John Chappell, say if he would be with us in the final photo up on the steps of the Pyramid Pointe?
>
> I can only mention some names who I knew and met in person and believe that they did a great job to make all this possible: John C. Chappell, Neil and Eleanor Munch, Dr. Cynthia Whitney, Dr. Domina Spencer, Francisco Miller, Ronald Hatch, R. H. Dishington, Milo Wolf, from fifteen years ago, and the present group of pertinacity, carrying the torch against the huge resistance of "occultism."
>
> Bob, Patricia and David de Hilster, Greg Volk, and many others who we only perceived as a busy persistent association of dissatisfied philosophers who will not rest until the last molecule of occultism disappears from natural philosophy.
>
> We—Steven and Rochelle—are close to ninety and could only be proud to be part of the NPA. As we'll see, soon, official science will realize where the solutions of centuries-old enigmas were coming from.
>
> *CONGRATULATIONS TO ALL OF US!*

Unintimidated, confident, swelling in number, members of the NPA present us with a curiously modern dilemma.

It is one that was summed up in an unrelated but wonderfully pertinent cartoon in the *New Yorker.* The cartoon, by Steiner, shows a dog sitting at a computer, vigorously typing away. Beside him is a canine companion to whom he explains, "On the internet no one knows you're a dog." In the age of the Internet, when anyone can set up a Web site and all of the attendant paraphernalia that digital technology can deliver—the archives and abstracts and database search functions, the videotaped lectures—how do we know, and how can we determine, who the "dogs" are?

Steiner's cartoon captures something fundamental about the zeitgeist of our time, the opening up and democratizing of knowledge, along with historically unprecedented opportunities to present the appearance of authority. The quandary we are faced with via the NPA is now that anyone can a publish a theory of physics online, what *can* be, or *will* be, or *should* be, our criteria for credibility in this field?

The questions I am asking here are not ones I have come to easily. I well remember the pride I felt each day as a university science student thirty years ago when walking into the physics building; it was as if there were a sign chiseled in the granite above the entranceway reading, ENTER YE INTO THE LAND OF TRUTH. We physics students *knew* that what we were studying was different from all that was going on in the humanities departments across the quad; *we* were studying eternal truths—the laws of nature—things that would be the same in a thousand years. We were encouraged in this confidence by our professors and by our own feelings of importance. When I began my career as a science journalist, I truly thought it was my mission to help nonscientists learn about a more fundamental realm of knowledge. But is this an accurate conception of what theoretical physics achieves? And if it is, then who gets to decide *which* theories are representative of this "deeper" level of "truth"?

To many people, myself included, such questions are unsettling. Of all the things in our society whose claims to objective truth have been reevaluated over the past century—religion, aesthetics, philosophy, and morality among them—physics has, until recently, seemed to stand apart. What are the consequences if anyone who feels like it can proclaim his or her own theory of physics? How are we to tell whose theories to listen to and which ones to take on board? As with so many other fields of human endeavor, the traditional method of adjudication has been credentials. We have trusted physicists with Ph.D.s and academic positions, which in turn have been vetted by other people with credentials so that standards have applied. The matter that now presents itself is whether or not theoretical physics is going to be one of those fields of endeavor that move "beyond" such formalized gatekeeping. With the rise of the NPA, I am suggesting that in some sense it already has. Whether or not NPA theories are taught at Harvard and Stanford, they are being presented to the public as genuine alternatives for what physics might say.

Credentials are by no means an insignificant issue, and as someone who has slogged through years of university science exams, I do not mean to discount them—no one who has participated in British-style academic training can be unmoved by the initials of a degree. Yet almost everyone agrees that credentials cannot be our only criteria for practice in many fields. Picasso did not have an MFA, and neither did Marcel Proust. Bill Gates did not graduate from college, and Einstein didn't have a Ph.D. when he published his first papers on special relativity. These are exceptions, perhaps, but the question arises as to whether credentials are necessary in any particular field. I think we might agree that when it comes to brain surgery, we would like our practitioners fully credentialed, yet when we look at medicine as a whole, the picture is not so clear. Americans now spend billions of dollars a year on alternative health practitioners, and many people feel they get tremendous therapeutic benefit and even lifesaving help from acupuncturists, herbalists, chiropractors, and the vast range of other non-MD health service providers.

Likewise, we do not care if painters or novelists have degrees. Or at least the art-loving and literature-reading public doesn't much care. Art and literature are interesting cases because these days, galleries and publishers *do* care. Increasingly, emerging artists cannot get into exhibitions or find gallery representation unless they have a Master of Fine Arts, while young novelists are often now expected to have graduated from a formal writing program. Credentials signal to the industry that a writer has passed through certain intellectual hurdles, yet it is an open question whether they are increasing our stock of literary merit. Mark McGurl, a professor of literature at UCLA, has written a fascinating book on the rise of degree writing programs in postwar America and the ways in which they have come to define the contemporary literary landscape, entitled *The Program*

Era: Postwar Fiction and the Rise of Creative Writing. Yet if it is true that publishers now care about qualifications, it is not clear that the state of letters is better for this development.

What happens to the republic of culture if only people with MFAs, at roughly a cost of $50,000 per degree, can hope to get published in major magazines or exhibited in major museums? One thing certain is that culture does not cease outside the academic sphere, and lots of people with no qualifications will carry on writing and making art despite being ignored by the mainstream. The whole category of "outsider art" or "folk art" operates independently of the credentialing art system and is growing rapidly in popularity. It has now spawned its own thriving tranche of magazines and galleries operating parallel to, and outside of, the mainstream art world. To my mind, outsider artists have produced some of the most powerful visual experiences of the past century. When I look at one of Martín Ramírez's hallucinatory images filled with trains tunneling through mountains, drawn on sheets of paper he cobbled together from scraps of newsprint welded with chewed-up bits of potato, I am thrilled by the vitality of his work. Ramírez was housed in a mental institution for much of his adult life; his personal story is heartbreaking, yet his art is breathtaking. "Where does this train go?" one curator asked. "It goes to drawing city."

That is a place to which no MFA can guarantee passage. Some individuals—credentialed and not—find their way to "drawing city," and some people, no matter how many degrees they hold, will never get to that enchanted place. Ramírez is now becoming recognized as a major artist of the twentieth century. A case has also been made for the Swiss outsider Adolf Wölfli and the Chicago janitor-draftsman Henry Darger, and

I would like to add Emma Kunz and Hilma af Klint to the list. None of these people went to art school, some of them were deemed mentally ill, all of them produced works of exceptional power. To which category does theoretical physics belong: Is it in a class with brain surgery, which functionally demands a degree, or is it more appropriately grouped with art and literature? In short, should theoretical physics be done only by those with recognized credentials? Or is it a field open to anyone who wants to have a go?

The recent trend toward credentialing in the arts represents a deviation from what has been a long and definitive cultural movement in the opposite direction. One of the hallmarks of modern Western society has been the gradual democratizing of the processes of artistic production. In the fifteenth century, most people could neither read nor write; today, anyone with a laptop and an Internet connection can write books and distribute them online. A similar transformation has occurred with music: In the eighteenth century, most people would never hear an orchestra play—not ever in their life—whereas today, anyone with a computer can download samples of the London Philharmonic and create symphonies. In three hundred years, we have gone from Bach composing fugues for a courtly elite to Beck producing records in his bedroom. Most of us applaud these egalitarian social transformations. We nod with approval as new technologies and new media make the power of authorship available to ever greater sectors of our society. We hail the openness and availability of knowledge and pride ourselves on living in a world in which participation has become an almost sacred principle. Why, then, should we draw a line at theoretical physics?

In a post-Enlightenment culture in which *access* and personal *engagement* have increasingly served as measures of progress, is there a point at which "progress" should stop? And even if you think it *should* stop, a question that now stands open is whether a movement that has already seized the opportunity *can* be stopped. Like so many other populist movements, outsider physics reflects needs and desires that are not being fulfilled through officially sanctioned channels, needs and desires that cannot be denied and refuse to be suppressed.

The democratizing tenor of the modern age is something of which NPA members are excessively aware. That academic physicists have resisted this trend is a source of continuing irritation. In "Our Minimum Consensus," Peter Marquardt rails against the "science lords" who hog theoretical physics for themselves and points to the NPA's inclusive and participatory spirit as a paradigm of resistance. According to Marquardt, the public also is hungering for a science that is more accessible. "The public, not specialized in physics, accepts [the NPA] approach readily because it matches with their everyday experience," Marquardt writes. "They find it rewarding to be invited to follow the scientific line of reasoning instead of being scared by weird ideas camouflaged with horrifying formulas." If there is one thing that unifies members of the NPA, it is a belief that the "formulas" of academic theoretical physics have become horrifyingly weird. By these lights, humanity's dialogue with the physical world has been hijacked by a group of experts who are trying to deny the rest of us participation in the conversation. Refusing to be intimidated by what they see as this elitist clique, members of the NPA make the claim that nature speaks a language the common person ought to be able to understand. In rejecting mathematics as the universe's tongue, the Natural Philosophy Alliance insists on a "scientific line of reasoning" that conforms to "everyday experience." Though they may not agree on how best to fix the problems of contemporary theoretical physics, members of the NPA are unified in their belief that by getting under the hood of the universe, they can free it from its maladies and rebuild it on literally more sense-able foundations.

Outsiders aren't the only ones who have worried about the mathematization of this science. In the middle of the nineteenth century, Michael Faraday expressed a similar concern. As a child of poverty who hadn't studied at a university, Faraday could not understand the mathematics into which his field ideas were ultimately transformed, and toward the end of his life he was upset that his beautiful concept had been rendered impenetrable to anyone but the mathematicians. Faraday died a hero, but an alien in the world he had helped to create; he could no more understand Maxwell's equations than the average paradoxer. Maxwell's friend William Thomson also had reservations. Although as we have seen, Thomson was a great admirer of mathematics and used it widely in his work, he insisted that behind the mathematics *physics* must always be driven by concrete *physical* ideas. That is why he was so enamored with vortex atoms: Vortex rings and vortex knots presented a clear picture of what it was the mathematics was supposed to be describing.

Yet the tendency of physics over the past century has been firmly in the direction of mathematical abstraction, a development that seems to both capture nature precisely and at the same time erase it. None of us will ever *see* the four-dimensional spacetime of general relativity or *touch* the eleven-dimensional manifold of string theory. These extraordinary concepts describe a world that shimmers like a mirage, exquisite and intangible and increasingly unreachable by the human body and mind. I hold degrees in physics and mathematics, and writing about these subjects is my profession, yet I struggle to understand general relativity. Any nonphysicist who says otherwise is not being honest. And so a tension arises in our society: If one of the purposes of science is to help us feel "at home in the universe" and to give us a sense of where we humans stand in a wider cosmological scheme, then how are we to respond when the world picture endorsed by the leading institutions of our society alienates large sectors of our community.

I have come to think of this as the "cosmological problem." Traditionally, the purpose of a cosmology was to embed a people in a world—what happens to a society when its official cosmology becomes one that 99 percent of its population does not understand and very likely cannot ever hope to comprehend?

I have cited the democratizing tendencies of artistic production in the modern era, but historically, one of the first great movements toward citizen engagement was in the sphere of theology, in the Protestant Reformation of the sixteenth century, and it is here, I think, that we find a particularly intriguing parallel with the NPA. Among the many innovations that Martin Luther introduced into Christian culture, one of the most revolutionary was his vernacular translation of the Bible. Prior to Luther, the Bible had been available only in Latin or ancient Greek and Aramaic. Luther insisted that this sacred text should be available in native languages as well so that all men and women could read it for themselves. Against the Roman Catholic Church's practice of setting priests as the mediators between God and men, Luther declared that every person should be able to study the word of God directly, without mediation. By making the Bible available in vernacular German, Luther offered a radical challenge to an intellectual elite that for hundreds of years had regarded knowledge of the Real as a specialized treasure to be accessed only by those with formal training. In contrast with the academic Latin-based theology that had characterized the Catholic Church for a thousand years, Luther wrote books whose aim was to make Christian thinking accessible to people in every station of life—to merchants and farmers and tradesmen, even to women and children. Through his belief that access to the Ultimate was the right of all human beings, Luther's approachable accounts of Christian philosophy helped to precipitate not only a theological revolution, but deep social change.

Where Luther insisted that the Book of God be open to everyone, so today members of the NPA insist that the book of nature can be read by us all. As the NPA presents it, nature is not the exclusive purview of the "science lords," but a treasure to be shared by all human beings. Like Martin Luther and the sixteenth-century critics of the Catholic Church, Jim Carter and the NPA are calling for a reformation, in this case a reformation of science. Their call is specifically for a new era in theoretical physics that would enable the participation of all willing minds. If it is extremely hard for those of us who have been trained in science to imagine what a citizen-driven physics movement might look like, we are stuck with the rather unexpected fact that with the NPA we already have one.

Chapter Eleven

SWIMMING PHYSICISTS

THE AMERICAN ARTIST Man Ray once remarked that there is no progress in art any more than there is progress in making love; there are only different ways of doing it. Science, however, is supposed to be all about progress as theories become more powerful, encompassing, and predictively accurate. To take just one example, consider the atom. One hundred fifty years ago, atoms were mere speculation, and until the end of the nineteenth century some very senior physicists denied they existed. Today, using scanning tunneling microscopes, we can see individual atoms and build nanoscale structures one atom at a time. Our understanding of atomic physics has led us to be able to synchronize atoms in such a way as to produce coherent waves of laser light powerful enough to reach the moon, and we can make atomic-scale corrals to trap a single photon. The most accurate clocks are now based on measurements of cesium atoms, so that time itself is registered by an atomic phenomenon, and with particle accelerators physicists have created more than a dozen new elements, extending the periodic table beyond the natural order.

Our understanding of atomic-scale physics and chemistry

has led scientists and engineers to create a huge range of supra-natural materials: semiconductors, superconductors, liquid crystals, and sheets of graphene one atom thick that are expected to revolutionize the way we use electrical power. From vague speculations in the seventeenth century, we have progressed to entire research programs dedicated to teasing out the potentialities of every kind of atom.

All of which makes it tempting to conclude that science does not belong in the category of art or making love, but represents a qualitatively different kind of enterprise. This has been the canonical view. Considering Man Ray's insight, an editorial in the British magazine *New Scientist* in 2010 stated that "from [this] perspective, science is a happier field of human endeavor. There are many different ways of doing science," the editors noted, "but do it right and you could change its course forever," the implication being that, unlike art or love making, science *is* defined by an axis of advance.

The "progress" view is allied with a metaphor deeply entrenched in many scientists' minds: Here science is seen as a mountain that generations of workers devote themselves to climbing. "Mount Improbable," Richard Dawkins has called it. Physics is one mountain, biology another, and so on, all incorporated in one magnificent range. "Progress" is equated with elevation up the slope, which is determined by the expansion of our "view." Thus Einstein ascended higher up the mountain of physics than Newton, because Einstein's theories of motion and gravity allow us to see all the terrain that Newton's revealed plus a good deal more. In theoretical physics today, general relativity and quantum theory appear to be two separate peaks, but they must actually be different parts of one larger peak because there is only one universe they describe. We already know places where the two intersect, and the goal of theoretical physics now is to ascend to a place from which it will be seen that both are local maxima on one theoretical Everest. Some people hope that all the peaks of science may one day be unified. As Stephen Hawking has written: "The eventual goal of science is to provide a single theory that describes the whole universe."

From a functional point of view, the progress view of science is extremely hard to argue with. Sitting here typing on a laptop loaded with blindingly fast microchips, I personally feel grateful to all those physicists who have worked out the details of quantum mechanics and who continue to advance its finer points, thereby making even better chips possible. When I listen to a car's navigation system guiding me to my destination, I understand that the GPS satellites tracking my whereabouts are factoring in timing corrections based on both special and general relativity. The combination of these effects means that clocks on board each satellite tick faster than equivalent clocks on the ground by thirty-eight microseconds per day. If this correction weren't taken into account, the ability of the satellites to fix a position would quickly deteriorate, so much so that positional error would accumulate at a rate of ten kilometers a day, making the system worthless. Without relativity's precise predictions, the global positioning system simply could not function. Practically speaking, only the most committed Luddite could deny that *some* kind of progress is being made, and when pressed, champions of science tend to justify the endeavor by its instrumental achievements.

Yet this is not the aspect of physics that captures public attention, nor is it the quality that turns physicists into celebrities.

Stephen Hawking, Steven Weinberg, Brian Greene, Leonard Susskind, Lee Smolin, Alan Guth, Sean Carroll, Lisa Randall—the most famous names in physics today—none of them produce theories applicable to daily life. Yet these are the people whose work receives enormous attention in the press and whose books are publishing industry events.

As a working journalist, I once tried to interest a major science magazine in an article about the physicist who had invented the magnetic coating that made read-write CDs possible, a technology that had benefited a good percentage of the people on earth. "Not interested," I was told. "Have you got any more stories about the Big Bang?" *New Scientist* frequently polls its readers, and surveys consistently show that what audiences want is the mythical stuff: the birth of the universe, the end of time, dark matter, dark energy, time travel, new dimensions of space, far-out ideas about gravity, possible new forces of nature, and other hypothesized universes. These are the stories that make it onto the cover of *Discover* and *Scientific American* and into the pages of the *New York Times* "Science" section. It is not the functional side of physics, but what I am choosing to call its *cosmological* dimension that has become by far the most valorized face of this science.

In the cosmological sense, "progress" in physics is generally deemed to be what advances us toward the summit of the physics mountain, what Stephen Hawking has called "a complete unified theory." But how exactly do we gauge progress of this kind? Many theoretical physicists have made it clear that practical applicability cannot be our yardstick here. Writing in *A Brief History of Time*, Hawking tells us that "the discovery of a complete unified theory . . . may not aid the survival of our species. It may not even affect our life-style." Because the "theories that we already have are sufficient to make accurate predictions in all but the most extreme situations"—inside black holes or in the first split second after the Big Bang, for example—a complete theory is not likely to advance any of our daily needs. Until very recently, theoretical physicists tended to tout the use*less*ness of their research.[1] In the cosmological domain of understanding, the "final theory" is widely regarded as its own reward irrespective of any application. But unlike the functional side of science—where a microchip either runs faster or it doesn't—it is often difficult to define what constitutes movement toward this goal.

"Ever since the dawn of civilization," Hawking writes, "people . . . have craved an understanding of the underlying order in the world. Today we still yearn to know why we are here and where we came from." Our desire to comprehend an "underlying order" may well be a justification for pursuing something, but the problem we now face is *which* ideas about such an order we are going to pursue. Tens of billions of dollars of particle accelerators and deep-space telescopes are currently being deployed in pursuit of this dream, and many more expensive instruments are being planned, so it has become no small matter of social interest how we choose to answer. To put the issue another way, we are faced with the question of *whose* concepts of a universal order we are going to invest in, or even consider taking seriously. For much of this book I have been presenting this dilemma as a lopsided battle between outsiders on the one hand and academic theoretical physicists on the other, yet the full extent of the situation is stranger—and ultimately more challenging to our ideas about how science functions, or at least how it is supposed to.

The "mountain" view of science takes for granted that as we

progress up the slope we remain on terra firma, grounding our theories in *material facts* that the community agrees on. Theoretical physics is supposed to be the bedrock of scientific ground, a deep stratum that in some sense forms a foundation for our understanding of the rest. For much of the history of modern science this idea has had a good deal of validity, but theoretical physicists themselves are now proceeding in a way that calls this view into question. In their current attempts to find a "final theory," academic theoretical physicists are becoming increasingly untethered from data. Their science is becoming insubstantial, losing its link to evidence, as if the top of the mountain itself is dissolving in a cloud of mist. The mist, moreover, is multivalent with almost every theorist proposing his or her own version of what a final theory might be.

At the NPA conference I attended in 2010, 120 different theories of the universe were on show. That may sound like a large number and we may smile with amusement at this fact, imagining the competing views and voices on the fringe, yet it pales into insignificance by comparison with the number of theories the insiders are now offering. In 2003, I attended a conference on the new field of "string cosmology," which extends the insights of string theory to the universe as a whole. String cosmology is a way of uniting general relativity with quantum theory that in many insiders' eyes represents our best hope for a complete theory of reality.

String cosmology is a protean field, and one of its virtues, if you choose to see it that way, is that it seems to come in a very large number of varieties. Estimates of how many versions of the theory there are keep going up. Today it is believed there are at least 10^{500} legitimate variants—that is 1 followed by *five hundred*

zeros: 100,000,000,000,000,000,000,000,000,000,000,000,000,00
0,000,000,000,000,000,000,000,000,000,000,000,000,000,000,00
0,000,000,000,000,000,000,000,000,000,000,000,000,000,000,00
0,000,000,000,000,000,000,000,000,000,000,000,000,000,000,00
0,000,000,000,000,000,000,000,000,000,000,000,000,000,000,00
0,000,000,000,000,000,000,000,000,000,000,000,000,000,000,00
0,000,000,000,000,000,000,000,000,000,000,000,000,000,000,00
0,000,000,000,000,000,000,000,000,000,000,000,000,000,000,00
0,000,000,000,000,000,000,000,000,000,000,000,000,000,000,00
0,000,000,000,000,000,000,000,000,000,000,000,000,000,000,00
0,000,000,000,000,000,000,000,000,000,000,000,000,000,000,00
0,000,000,000,000,000,000,000,000. It is hard to overstate the enormity of this number. To give some perspective here, physicists estimate that our universe contains 10^{80} subatomic particles—that is, 1 followed by eighty zeros. Thus the number of possible string cosmologies exceeds the number of particles in our universe by more than four hundred orders of magnitude! There is nothing in the physical world to which we might compare this unimaginably vast quantity. Computer scientists and small children sometimes use a googol, which is 10^{100}, as an example of a really huge number, yet the number of possible string theories is a googol *squared*, then *squared* again. Every one of these theories describes a different possible universe with different physical laws, and every one of them falls under the umbrella of the theoretical physics mainstream. As far as insiders are concerned, competition for the ultimate theory comes not from outsiders, but from colleagues down the hall.

That string cosmology conference I attended was by far the most surreal physics event I have been to, more bizarre than any NPA event for the very reason that this was not a fringe affair but a star-studded proceeding involving some of the most famous names in science. Stephen Hawking was present, along with most of the leaders of the field: string theory pioneer Leonard Susskind of Stanford; Harvard's Lisa Randall, and Princeton's Paul Steinhardt, architect of the thrillingly titled "ekpyrotic theory," a string cosmology variant that proposes that our universe cycles through an endless series of big bangs and "big crunches," exploding, then collapsing and exploding again. On the conference program also was Brian Greene, string theory's most public champion and the author of several hugely successful books on the subject, including *The Elegant Universe.* At the time the conference was being held, PBS was launching the television series of the book and Greene was off doing publicity; he seemed to be all over the news. Every time you turned on the TV, there he was. In short, the timing couldn't have been more perfect.

The meeting itself took place at the University of California at Santa Barbara, at the prestigious Kavli Institute for Theoretical Physics, whose director, David Gross, would the following year be awarded the Nobel Prize. Outside the institute the California sun sparkled on the ocean, while inside another kind of pyrotechnics was in play. Throughout the three-day meeting, each speaker delivered his or her version of a string-based cosmology, outlining a conception of what the universe might look like from an ultimate point of view. Few were content with a single universe. What had fired many of them up was an idea proposed by Susskind, according to which every possible variant of string theory represents a real physically existing universe, billions upon billions of which might *coexist* in some superuniversal space. "Landscape" was the term Susskind had coined to describe all the possible *theories*—the mathematical totality of the system. Any array of actual universes was known as a "multiverse," and at the Santa Barbara conference, speakers presented their ideas about different possible multiverses.

Speaker after speaker offered PowerPoint presentations describing how different arrays of universes might foam in and out of existence. What exactly this cosmic "population" looked like differed for each person. Some thought each universe was born free, while others believed the multiverse would grow organically as new universes spawned out of previous ones. Some posited that the multiverse had existed forever, while others thought it had come into being at a specific time—though what exactly "time" meant when there is more than one universe was hard to say. Some thought the multiverse was indestructible, while others were convinced it would one day end. There were "toy universes" and "baby universes," and some people allowed that there might be constraints on the possible sets of universes. For others, the limits were set only by our imaginations.

Over tea break on the final day of the conference, I chatted with professor Joe Polchinski, who had organized the event. Polchinski was the person who'd been keeping track of the number of possible string theories, and was also responsible for making sure that speakers stuck to the schedule. Not even Hawking was allowed to exceed his allotted time at the podium. As we sipped tea, I asked Dr. Polchinski what he had thought of one speaker who to my mind had delivered a particularly dazzlingly talk. "Utterly splendid," he replied. Or words to that effect. "Of course there's not a shred of evidence for anything the fellow said," he went on.

It didn't seem to matter whom I spoke to, or who the speaker was, that was the response I got. Whoever I talked with assured me that everybody else's theories were unsupported by evidence and based entirely on arbitrary assumptions. None of this was driven by physical discoveries; it was all being inspired by possibilities inherent in the math.

Nobody seemed perturbed by this extraordinary state of affairs, and after the tea break they plunged back in, eager to hear about yet more of the Landscape of possibility. With 10^{500} varieties of theory to choose from, there was no lack of options, and there seemed to be no limits either on sheer imaginative zeal. In just thirty minutes at the podium, each of these brilliant men and women served up a phantasmagoria in which universes exploded like fireworks in wondrous arrays. Everybody started with different assumptions—a different set of "initial conditions"—and proceeded to watch as their own particular multiverse unfolded from the equations. Many of them had simulated their ideas on computers, and some had animations to show. There were charts and diagrams and statistics about various possible universe populations, and in the near infinitude of options, there didn't seem to be much that wasn't in the realm of probability. As at any NPA event, everyone had the Answer and could barely contain their excitement about sharing it with their peers. Sizzling with enthusiasm, inspiring one another with their mathematical flair, most of them spoke at a million words a minute, trying to pack in the maximum number of ideas. After two days, I couldn't decide if the atmosphere was more like a sugar-fueled children's birthday party or the Mad Hatter's tea party—in either case, everyone was high.

For sheer bizarreness, the account of reality given to us by string theory rivals anything the NPA has to offer. If it weren't coming from the academy, one could be forgiven for thinking this was a new kind of fairy tale. According to string theory, everything in our universe is made up from minute knots of some infinitely thin, stringy stuff. Different particles of matter are different configurations of this fundamental stuff, each one vibrating like a miniature rubber band. Depending on which version of the theory you favor, our universe has either ten or eleven dimensions. (In some especially mysterious versions there are twenty-six dimensions.) This means that in addition to *time* and the three dimensions of space that we humans experience directly, there are supposedly another six or seven spatial dimensions we do not see.[2]

Where are these extra dimensions hiding? According to string theory, they are curled up in a ball so tiny, we cannot detect them with our usual measuring tools. They operate on a scale many orders of magnitude smaller than the subatomic particles, and one of the goals of string theory is to chart the structure of this micro-microscopic world. String theory describes its additional dimensions through the mathematics of topology and "group theory," and the reason there are so many variations is that there are lots of different ways in which six or seven dimensions might be configured. The strings that inhabit this miniature space are sometimes described as immensely complicated knots, so that string theory realizes the vision of matter that Tait and Thomson dreamed of in the nineteenth century—although the mathematics is now much more complex.

The way string theory is usually explained is by analogy with a hose. Imagine that you are an ant living on the outside of a long and very thin garden hose. Running along your hose each day, you think you are living on a line, which is a one-dimensional space. Then along comes a brilliant ant physicist who realizes that the line is *really* a tube, so your world has a second, hidden dimension. The shape of this dimension is the cross section of the hose—in other words, a circle. Most ants don't perceive the two-dimensional nature of their world because the hose is so thin that the second dimension bypasses their awareness. Only the ant physicists with their massively powerful ant accelerators can detect it. According to string theory, we are like these ants: We live in a world that has more dimensions than we can perceive with our senses. Only *our* physicists armed with massive particle accelerators can hope to detect these additional dimensions of being. The specific configuration of this tiny hidden space—the cross section of *our* multidimensional "hose"—is one of the things string theory describes.

Like Tait and Thomson's vortex atom theory, string theory addresses the question of *what matter is* by proposing an essentially spatial answer. According to Thomson, "matter" is knots in the "ether." But that left open the question of what the "ether" is. Tait's answer was to propose ever more layers of ether filled with ever smaller knots, each layer constituting another space of being. String theory stops this infinite regression by proposing just one space of being, but that space itself has extra dimensions. To the question "What is matter?" the string theorist replies: "It is vibrations in a multidimensional space." The price we pay for stopping Tait's infinite cascade is a description of reality that almost no one can comprehend. To most mere mortals the equations of string theory may as well be written in Martian, and there are no more than a couple of thousand people who understand them. For Jim Carter and other physics outsiders, string theory has replaced quantum theory as the paradigmatic case of insiders gone berserk. "Every time string theorists can't explain something," Jim quips, "they add another dimension." To him, nothing seems more outrageous than extra dimensions of space we cannot hope to ever experience ourselves.

String theory is the apotheosis of Tait and Thomson's vision, but its roots can also be traced to a man who occupies a uniquely strange place in the history of physics. A brilliant mathematician and maverick genius, Theodor Kaluza became, in 1919, the first person to propose that our universe might contain more dimensions than four. German by birth and a near contemporary of Albert Einstein, Kaluza grew up as the son of a languages professor and throughout his life he effortlessly learned new languages, delighting in the play of ideas that other tongues allowed. His favorite written language was said to be Arabic and it is interesting to speculate that in the arcs and curlicues of this elegant script he might have found inspiration for the mathematical embellishments he would soon propose.

Kaluza's proposal to extend the number of natural dimensions arose out of his interest in general relativity, a theory Einstein had published in 1916. Just three years later, Kaluza wrote to his countryman to suggest that the theory might be enlarged to five dimensions. In addition to time and the three usual spatial dimensions, he suggested adding another dimension of space. His reason for doing so was surprising: It turned out that when he wrote Einstein's equations in five dimensions, the higher-dimensional version of the theory contained within itself not only general relativity but Maxwell's equations of electromagnetism. Looked at in five dimensions, these two seminal theories became complimentary parts of a wondrously elegant whole.

Einstein was not usually one to balk at innovation—that was supposed to be his specialty—yet the prospect of an unseen dimension alarmed him, and he rejected Kaluza's idea. Like Jim, Einstein had a fondness for tangible things. He had trained as an engineer and worked at a patent office; he believed in physical reality. Kaluza's extra dimension seemed to have no grounding in the real world. But the elegance of the mathematics tugged at Einstein's consciousness and would not let him go. If beauty was indeed truth, then it was hard to resist the amazing grace of

Figure 13. Theodor Kaluza (SPL/Photo Researchers, Inc.)

Kaluza's equations. Einstein became intrigued. Where, he asked, did Kaluza think the additional dimension might be? Kaluza's response was the explanation about the hose. Einstein came around, and in the 1920s five-dimensional theory was briefly fashionable in the world of academic physics. Kaluza and the physicist Oskar Klein calculated the scale of the extra dimension, which turned out to be many orders of magnitude smaller than an atom. At the time there was no way scientists could imagine accessing such a tiny scale, and physicists turned away. With no means of verification, the idea seemed little more than a fantasy.

Kaluza himself was not so easily deterred. At the time he wrote to Einstein he did not have a professorship, but he was an outstanding mathematician with confidence in his powers. He was also very willful. One of the reasons his career was languishing and he hadn't been offered a proper job was that he was resisting the ideology of National Socialism. Academics were expected to buckle under and support the Nazis. Kaluza refused. If he could stand up to the rising forces of totalitarianism, he could certainly withstand a little scientific skepticism. He believed in his fifth dimension, he could feel it in his bones. He *knew* it was real and was convinced he could demonstrate its existence. If he couldn't test for it directly, he would test for it in principle. So he started reading up about swimming. Kaluza could not swim—that was the point: He decided he would read all he could about the theory of swimming, then once he had mastered the theory he would test out this knowledge in practice. Alone in his study, he plowed through calisthenics manuals, imbibing the act of swimming in his mind, imagining what it might be like to actually swim, and mentally putting his body through the motions. After months of preparation, the appointed day arrived and Kaluza escorted his family to the seaside, whereupon he hurled himself into the waves. Lo and behold, he could swim! In Kaluza's mind, the link between theory and reality was confirmed—his fifth dimension was real.

String theory today is an extension of Kaluza's work. The general class of multidimensional physics theories are now known as Kaluza-Klein theories. (There are variants other than string theory.) Kaluza had to deal with only *one* extra dimension because during his lifetime physicists knew only about one other fundamental force of nature in addition to gravity—the electromagnetic force. Since then they have discovered two other fundamental forces that operate inside the atomic nucleus. To explain all four forces in a Kaluzian fashion, the equations seem to demand an extra six or seven dimensions (and in some people's versions, an extra twenty-two). Whatever the final number, one thing hasn't changed since Kaluza's time: There is not a shred of evidence for any of them, unless of course you are prepared to count swimming physicists. String theory remains so far a purely mental exercise, a real-life incarnation of what the French absurdist writer Alfred Jarry called "pataphysics"—to wit, "a science of imaginary solutions."

The fact that the extra dimensions of string theory have not been verified has begun to cause some consternation in theoretical physics circles. Several prominent physicists who are not string theorists have written books denouncing it. Lee Smolin, the theory's most eloquent detractor, has wittily declared that with so many variations to choose from, string theory is not so much a "Theory of Everything" as a "Theory of Anything."

String theorists themselves are desperate to find some proof of their ideas, and are hoping that the new particle accelerator at CERN in Europe will throw some evidence their way. The Large Hadron Collider (LHC), as it is called, is the most powerful accelerator ever built; and at a cost of $9 billion, certainly one of the most expensive scientific instruments in the history of the world. Its energy is focused in a ring-shaped tunnel seventeen miles in circumference that accelerates two beams of protons and antiprotons to nearly the speed of light. The LHC was not designed specifically to test string theory, but its ring-shaped beams now carry the concentrated aspirations of the string physics community. Some evidence—*any* evidence—is needed if the enterprise is to stay float.

At the Santa Barbara conference, excitement among the assembled theorists was focused on a development that had recently been proposed. Some of them, not satisfied with being restricted to immeasurably tiny dimensions, had found a way in which the equations seemed to allow for the addition of a *huge* extra dimension, one so huge that our universe could sit within it. This gigantic space was dubbed the "bulk," while the space of our universe was dubbed a "brane," short for membrane. What was particularly intriguing was that many versions of this new "brane cosmology" allowed for the possibility that our universe was just one of a collection of universes coexisting in the bulk. Equally exciting was the prospect that this bulk-and-brane concept might be testable. In particular we were told that the Large Hadron Collider might be able to detect evidence of a large fifth dimension through anomalies in the behavior of gravity, which was said to "leak away" from the membrane of our universe into the infinitude of the bulk. As this book goes to press, string theorists are eagerly awaiting results.

The night after the Santa Barbara conference ended, I drove home to Los Angeles. It is a long drive, and I had plenty of time to think. I left Santa Barbara as the sun was setting, and for the first half hour the sky was a psychedelic mix of pinks and oranges that in my mind seemed a complement to the spectacle I'd been witnessing for the past few days. As darkness took over, the sunset was replaced by an equally mesmerizing display of taillights on the Pacific Coast Highway—red lights in one direction, white lights in the other, two endless streams curving and arcing with the line of the freeway. As I zoomed through this slipstream, my mind was reeling from what I had observed. I had been trained as a physicist in the late 1970s, and empirical verification had been drummed into us students as the hallmark of our science. I remembered reading about Stephen Hawking and Roger Penrose's work in which they had shown that black holes were physically possible. Although Einstein had thought black holes were an artifact of his equations that would never be realized in nature, Hawking and Penrose showed that at least in principle they were possible. It was dizzying stuff, but we students knew we must wait for evidence before we deemed them real. It was one thing to know that something wasn't disallowed by the equations, quite another to believe it was out there in the world. In Santa Barbara this empiricist caution had been thrown to the wind; the spirit of logical positivism was nowhere in sight. On the contrary, the attitude among the string cosmologists seemed to be that anything that wasn't logically disallowed *must be out there somewhere.* Even things that weren't allowed couldn't be ruled out, because you never knew when the laws of nature might be bent or overruled. This wasn't students fantasizing in some late night beer-fueled frenzy, it was the leaders of theoretical physics speaking at one of the most prestigious university campuses in the world.

Immersed in the river of lights on the Pacific Coast Highway, I thought about the images I'd been watching in Santa Barbara. All those universes swimming across the physicists' screens blended in my brain with the cars streaming around me. Our cars, I thought, are little "island universes," spaces we have to ourselves. Was this not one of the appeals of the car, that we each get to travel in our own automotive bubble? With the billions of universes offered up by string cosmology, I realized, we could each have our own spacetime bubble—indeed, a multiverse of our own. With its nearly infinite Landscape of possibility, string theory offers us the cosmological version of consumer choice.

Since that 2003 conference, string theory has turned out to have a truly extraordinary range of variations: There is heterotic string theory, superstring theory, and something called M-theory, an even more mathematically challenging expansion. There are now reckoned to be five major classes of string theory and strings themselves may operate at the atomic or cosmological scale. The astounding number of variations of the theory has become for some physicists a source of angst and for others a cause for celebration, for in this endless cascade of possibility just about *any* kind of universe can be encompassed by the "legitimate" laws of science.

In 2005, Leonard Susskind published a book called *The Cosmic Landscape* in which he made the case for embracing this plethora: "The Landscape is a dreamscape," he wrote. According to Susskind, in one part of the Landscape universes might have four dimensions, in other parts they may have five, or six, or seven dimensions. In one part of the Landscape, everything may be made up of strings. In another part everything may be composed of membranes. In yet another everything may be made from miniature black holes. Instead of trying to determine which one of these possibilities represents the True universe, Susskind advocated accepting them all as part and parcel of a wider world that includes *everything* the mathematics can describe. What we are being offered here, as Susskind put it, is "a Landscape of possibilities populated by a megaverse of actualities."

In a new book published in 2011, titled *The Hidden Reality*, Brian Greene also argued for taking the Landscape of string theory as literal physical truth. Greene noted that quantum theory also can be read as supporting the idea that all possible universes exist. Although this interpretation was first proposed fifty years ago, it is only now, with the support of string theory, becoming a mainstream academic position. If one takes the "many worlds" interpretation of quantum theory at face value then not only do other universes with other physical laws exist, but every possible version of our *own* universe exists as well, somewhere in the multiverse. Thus, according to Green, the multiverse contains endless near-facsimiles of *this* world, in each of which historical events have played out rather differently. In one of these "parallel universes" you are reading this book and scratching your head in puzzlement with your prehensile tail.[3]

The Landscape of string theory and the many-worlds interpretation of quantum theory have combined to produce an extraordinary development in the rhetoric of science. The metaphor of a mountain representing one Final Theory is now being replaced by the idea that every possible theory is just one more peak in an infinite topography of being. In this way of seeing, the search for an "ultimate" theory becomes moot, for every universe that might happen is actually here.

Powering along on the Pacific Coast Highway on the way home from the Santa Barbara conference, I had to admit to myself that I had barely comprehended what many of the string cosmologists had been saying. In truth, most physicists don't understand this stuff either. I recalled a line from Flann O'Brien's surrealist novel *The Third Policeman*: After one of the characters in the novel has atomic theory explained to him, he responds: "What you say must surely be the handiwork of wisdom for not one word of it do I understand." I was certain that in Santa Barbara I had heard "the handiwork of wisdom." What it added up to was difficult to say.

A further literary reference came to mind, Alice's remark after she reads the poem "Jabberwocky": "Somehow it seems to fill my head with ideas—only I don't exactly know what they are!" I understood how she felt. My head also was filled with ideas I didn't know how to parse. Shimmering through the lights on the highway, I seemed to be in a dream, the trancelike threads of the cars mixing with the trails of all the simulated universes I'd been shown.

Was any of it real? Did any of it exist? Or was it all an artifact of the mathematical imagination?

Lewis Carroll would, I think, have appreciated where we stand. His *Alice* stories were inspired in part by seemingly absurdist developments in mathematics. Carroll intuited, as few of his colleagues were perhaps able to appreciate in the nineteenth century, the imaginative potential that was being unleashed by the new language coming into being through innovations in algebra, geometry, and topology. "Jabberwocky" is not a reflection on mathematics per se, yet it seemed to me that this cryptic epic might well serve as a parable for what physics had become. "Jabberwocky" is sometimes said to be the greatest piece of nonsense written in the English language; its individual words are made up, yet somehow they blend into a narrative that we instinctively understand:

'Twas brillig, and the slithy toves
Did gyre and gimble in the wabe;
All mimsy were the borogoves,
And the mome raths outgrabe.

"Beware the Jabberwock, my son!
The jaws that bite, the claws that catch!
Beware the Jubjub bird, and shun
The frumious Bandersnatch!"

And so on.

I realized on that drive home to Los Angeles that in Santa Barbara I had experienced similar, semiconscious sensations: "A Jabberwock, with eyes aflame, came whiffling through the tulgey wood" at me. It "burbled as it came." I may have been bamboozled by what I had seen, but I was also electrified. Indeed, I "chortled" in my joy! On that long, light-streamed journey, I realized that the physics of my heroes—Newton and Kepler and Einstein among them—had effectively become a new form of storytelling, spawning a new genre of speculative imaginative literature. As with "Jabberwocky," it did not seem to matter if we understood the words physicists used or if the authors made them up. It did not seem to matter if branes and bulks and extra dimensions of space "actually" existed. Did it really matter, I wondered, if physicists proposed ten universes or ten gazillion? We could all be carried along by the power of their narratives and engulfed in these bizarre magical worlds. As with "Jabberwocky," it seemed to me that in some way string theorists had reached a point of making sense while simultaneously transcending it. I determined then that I wanted to write a book about outsider physicists that would serve as both a response to, and in some sense as a reflection of, what the insiders were doing.

The French novelist Marcel Proust had an insight that seems germane to my subject. Proust once said that "great writers invent a new language within language, but in such a way that language in its entirety is pushed to its limit or its own 'outside.'" With extra dimensions of space, topological knots of string, and all the other enchanted paraphernalia of their field, string theorists have invented a new language, in the process propelling the language of science to *its* outside. In the age of string cosmology, we seem to have arrived in a new kind of Wonderland in which the empirical verities that Francis Bacon championed have been replaced by the values of fiction. To quote Jim

Carter, who in turn is quoting Einstein, in the upper echelons of the theoretical physics establishment "imagination *has* become more important than knowledge."

If outsiders also appear to be "making things up," might we not simply enjoy their alternative narrative arcs?

Afterword

TREE RINGS

IN THE ARC of Jim's life, circlons have a habit of appearing in unexpected ways, a point that may be witnessed in the forests around Enumclaw and in the vicinity of the Green River Gorge. For many years now Jim has been planting redwood trees in the bare patches where loggers have clear-cut the firs. Sometimes he puts them along the roads so that as they grow they will enhance the view for motorists; other times he puts them out of common sight deep within the forest. He has planted the trees as far away as Seattle, though most of them are located in King County in and around the gorge. Jim buys the trees in California at his own expense. Periodically, he and Linda make a journey from Washington in which they drive down the coast road, through Oregon to Northern California, and fill the flatbed of the truck with young redwoods, then drive back home. Depending on the time of year, there may be up to a thousand baby redwoods resting in the Carters' yard.

One spring day, in a year like any other, I joined Jim for a planting excursion. His aim was to get about twenty trees in the ground in a patch of clear-cut a mile or so from the gorge. The day was crisp and incredibly fresh, one of those perfect

Washington afternoons when the skies are blue and dotted with fluffy white clouds. It wasn't raining, which in my experience was a miracle. Jim loves the rain, and when he lived in California he used to miss it, he says; the rain was one of the things that drew him back to Enumclaw. For Jim there can never be too much rain or too much forest, though even he will allow that when you're planting *more* forest, it is preferable to do so when it isn't raining. Fortunately it wasn't as we loaded up the truck that day with seedlings.

As we motored along the back roads of Enumclaw toward Mt. Rainier, the scenery switched between thick walls of forest and patches of clear-cut where the land looked violated. Behind the wheel Jim was in his element: This was the country that was in his blood and to which he had returned as an adult so that his children also might grow up here. His boys, Paul and Eric, were married now and having children of their own, and a new generation of Carters would have King County in their blood. Jim was in an expansive mood, and as we passed through a massive clear-cut he opened up about his forestry philosophy: "I've planted a lot more trees than I've cut down," he began.

> Whenever you see a clear-cut, the logging companies are required to go in and plant more trees. They usually just plant Douglas firs, but I'll go in there and walk around and put in a few dozen redwoods, particularly along the roads. I'm getting more and more trees every year. Last year I planted around three hundred. This year I got a thousand. Maybe next year I'll get three thousand. I plant them along the freeways and in places in Seattle. Wherever there's a nice view and it looks like a redwood tree would go there, I plant them. Usually I take a

few redwood seedlings with me, and whenever I spot a place where I think a redwood should be I'll put one.

In and around Enumclaw, the thousands of redwoods Jim has planted are gathering strength. Long after he is dead they will rise above the canopy of firs as a living gift to the world. This act of biotic enhancement is not something Jim makes a fuss about. It's a service he renders quietly, out of his love for the redwoods and the land, and because for all his interiority and perhaps a touch of Asperger's, Jim understands that the human web is what we collectively make it. Just as circlons bind together to make the patterns that give structure to matter, so we humans bind together to form the familial patterns that give structure to our lives.

For a long time Jim planted the redwoods randomly; then a few years ago a new idea occurred to him, and deep in the forest he began planting the trees in circles. These redwood rings are located secretly, away from the roads so they can grow undisturbed. In hundreds of years' time, these living *circlons* will thrust their gigantic bows high above the forest, their crowns fused, their trunks grown close together.

Jim has planted a small version of one of these redwood rings in a corner of his property near his house. After we got home from planting, he took me over to see it. As the trees grow up, he imagines the ring becoming a cubby-house for his grandchildren, and *their* grandchildren, to play in. Already a visitor can begin to sense what a closed ring will feel like, the magic of being encompassed in a total arboreal embrace.

But there is also something else here. Standing in that living ring, I realized Jim's whole life has been marked by a kind of organic excess. So much of what he does is, strictly speaking, unneeded. As with these redwood rings, Jim's projects have been compelled by his own inner drives and by a continuing thread of creative surfeit that points us—as art and literature and poetry can—toward possibilities outside the expected order.

Jim's redwood circlons have another role, too, for they point us to the physics he has spent his life developing. In these emblems of his theory Jim sees a sign, and perhaps a validation, of the theory itself. At all levels of our universe, nature finds expression in the circlon form—in the tracks of particles deep inside accelerators, in the giant plumes of ionized gas that shoot out from the surface of the sun, in galactic nebulae, coral atolls, and smoke rings, but most of all in the constituents of matter that Jim believes are the building blocks of everything. In the final analysis, Jim's circlon-shaped particles may also be seen as manifestations of a string theory, for his subatomic springs are also coils of some minutely thin "stuff." Jim came to the string concept more than thirty years ago, and the fact that this idea is now being embraced by the mainstream suggests to him that his other ideas will one day be vindicated, too.

Standing beside his redwood ring in the twilight of that lovely spring day, Jim mused about the totality of his work. As the sun set over the lawn where smoke rings had played, he pondered how citizens of the future may interpret the symbols he has left for them. "People might wonder about who planted them and about what they were supposed to mean," he said. Perhaps such wonder will inspire our ancestors to find out about the man responsible and to study the theory of the universe his trees have grown to represent.

This book is published in November 2011.

In 2012, Jim Carter will have been working on his theory of physics for fifty years.

APPENDIX

The Principles of Circlon Synchronicity and the Living Universe by James Carter

The fundamental assumption of Circlon Synchronicity is that protons and electrons exist in the universe and that they are exactly what we measure them to be. Circlon Synchronicity is a conceptual model of mass, space, time and gravity that is based on seven principles of measurement that describe the interactions of two fundamental particles of matter within the inert and featureless and infinite void of imaginary space. Within the standard model of physics, each of these principles is replaced with a metaphysical assumption that can't be measured.

Measurements show electrons and protons are the only absolutely stable and eternal primary particles of matter and that they both have a circlon shape. Photons and neutrinos are the only stable secondary fundamental particles of matter and are made up from parts of electrons and protons. A photon is a matter/antimatter pair made from equal pieces of an electron and proton. A photon can exist as a linear mass particle moving at the speed of light or as a circlon shaped mass particle forming a mechanical link between a proton and electron within an atom. A neutrino/antineutrino pair breaks off from an atom when a neutron is formed from a proton and electron. A neutrino is a piece of a proton and an antineutrino is a piece of an electron. All four of these basic particles are complex coil structures composed of cosmic string that has mass and is wound into circlon shapes. The whole particle zoo, both stable and unstable, from the neutron to the uranium atom, are all composites of two or more of these four fundamental particles.

The Seven Principles of Matter

Circlon Synchronicity is based on seven principles of experimental measurement that reveal the circlon shape of the structure of our reality.

1. THE PRINCIPLE OF ABSOLUTE PHOTON REST

Measurements of Doppler shifts within the vast numbers of the 2.7° Cosmic Blackbody Radiation (CBR) photons show that our solar system is moving relative to a common position of rest that is shared by all photons. All photons move at C relative to this *photon rest.* All photons move at (C +/− V) relative to moving bodies and these relative motions can be measured with Doppler shifts.

2. THE PRINCIPLE OF THE ABSOLUTE MOTION OF MASS

High speed measurements demonstrating the Lorentz Transformation show that a body's mass is increased when it is accelerated and then decreased to an absolute minimum value when it is decelerated to a position of photon rest. Careful measurements with atomic clocks show that the absolute values for mass change caused by either a body's inertial motion or its gravitational motion can be determined quite accurately. Clocks are slowed by the conservation of angular momentum as their mass is increased by absolute motion.

3. THE PRINCIPLE OF RELATIVE GRAVITATIONAL MOTION

Measurements show that the increasing dimensions of large bodies of matter produce a three-dimensional outward acceleration at their surfaces. This measurement is actually a deceleration that slows the surface of the body to a constant three-dimensional upward surface velocity. The constant for gravitational force is measured to be the outward velocity of 9.2116×10^{-14} m/s at the Bohr Radius of the hydrogen atom. All of gravity's phenomena can be explained in terms of these measurements.

4. THE PRINCIPLE OF THE CIRCLON SHAPE

The "circlon" is a precise and very complex triple torus shape that for most purposes can be visualized as a hollow doughnut. Measurements of protons, electrons and hydrogen atoms show that their parameters of mass, energy, wavelength, Bohr radius, fine structure, and radiation spectra require circlon-shaped mechanical particles. Atoms are formed when electrons and protons are held together by the dynamic motions of their circlon shapes. Measurements of the 282 stable atomic nuclear isotopes demonstrate that all of these nuclear structures can be physically assembled from the circlon shapes of the protons and neutrons.

5. THE PRINCIPLE OF PHOTON MASS

All experimental measurements of photons can be used to determine that they have both a shape and a mass. A photon's kinetic energy is $E = MC^2/2 + I\omega^2/2 = MC^2$, where "$\omega$" is the angular velocity and "I" is the moment of inertia. Its momentum, "p," is given by the equation $p = MC$. Its wavelength is $\lambda = 2\pi I\omega/MC$. Its angular momentum is $I\omega = M\lambda C/2\pi$ and this is the same value for all photons. A photon with *mass* means that there is never a transformation between mass and energy. Both the mass and the energy in the universe are eternal and remain constant and absolutely conserved. Mass and energy are complementary and inseparable components of both matter and photons that coexist like the two sides of a coin. Kinetic energy $E = MV^2/2$ is merely a measure of the motion of mass and does not ever exist separate from mass. The photon is not an "energy particle." It is a particle of mass and its energy is equal components of the kinetic energy of the linear motion of its mass and the rotational kinetic energy of the spin of its mass.

6. THE PRINCIPLE OF THE DUALITY OF TIME

Experimental measurements with atomic clocks show that there are two distinct and opposite flows of time. *Gravitational time* is the very slow absolute time based on the flow of gravitational motion. *Inertial time* is the very fast relative time based on the velocity of mass relative to the speed of light. Gravity clocks and inertial clocks run at the same rate at rest, but at *high* velocity gravity clocks run faster and inertial clocks run slower. The interchange between these complementary time flows can be demonstrated by the variations in very accurate clocks put into different orbits. Inertial time is slowed by both orbital velocity and gravitational surface velocity. In the low space station orbit, clocks run slower than they do on earth and in the much higher GPS orbit they run faster than earth clocks.

7. THE PRINCIPLE OF ELECTRON MASS AND SIZE TRANSFORMATION

Gravitational time is itself a duality between the slightly different gravitational motion rates of electrons and protons. Measurements deep into the cosmos show that photons emitted by atoms in the distant past have much longer wavelengths than the photons emitted by those same atoms today. The cause of this "redshift" phenomenon is that the electron has been gradually growing in size and losing mass over cosmological time. In the past, these more massive electrons caused atoms to emit their photon spectra with longer wavelengths. The complete evolution of the universe can be demonstrated by extrapolating the changing mass of the electron back to that point in the past when the masses of the electron and proton were equal. Using the circlon shape as a template, it is possible to calculate the exact point in cosmological time when the 2.7° Kelvin CBR was formed.

The Evolution of Matter in the Living Universe

My theory of the evolution of matter in the Living Universe is in strict adherence to the standard laws of physics and the dynamics of quantum mechanics. The only deviation from standard physics is the assumption that the mass ratio between the proton and electron has been slowly but constantly changing over time. It is this gradual decrease in the electron's mass that is used to explain how the properties of matter have slowly evolved over cosmological time. As we go farther and farther into the distant past, electrons become heavier and heavier. This causes the hydrogen atom to emit photons of longer and longer wavelengths. This is verified by the equations of quantum electrodynamics but it can also be demonstrated with pure logic.

In my theory the universe evolved to the tune of a single evolving "constant of nature." As the mass of the electron slowly decreased, the values of most of the other fundamental constants also changed in response. The *Bohr radius* got smaller and the *fine structure constant* got larger as the *electron mass* decreased. These changes, in turn, caused the photon *wavelengths of atomic spectra* to decrease and the *energy of atomic spectra* to increase. This caused the light from distant galaxies to appear to be Doppler shifted. The *speed of light* and *Planck's constant* did not change during the evolution of matter.

THE ANTI-HYDROGEN ATOM

The evolution of the Living Universe began with two particles, a positron and an antiproton that were coupled together to form an atom of anti-hydrogen. This single atom was as massive as the entire universe today. I have no explanation for the origin of this original atom. A devotee of quantum mechanics and the Big Bang might say that it just popped out through a tear in the fabric of the space time continuum. A religious person might say that it was made by god and an atheist might say that it was god. I am just making the metaphysical assumption that this anti-hydrogen atom simply existed at the beginning of our universe's evolution. This does not seem to be a large stretch of the imagination since today's universe is still basically just a bunch of hydrogen atoms.

Regardless of its origin, it was the negatively charged antiproton that was gradually decreasing in mass and increasing in size within the antiatom. As the antiproton lost more and more mass, it eventually reached a point where it captured the positron to form an antineutron. Within this antineutron, the mass of the antiproton continued to decrease until it finally became equal to the mass of the positron.

ANTIMATTER

A unique event occurred in the history of the Living Universe that had not happened before or since. The antineutron became a matter/antimatter pair that was confined at rest within the structure of what had just been an antineutron. The effect of this transformation was to make the positron behave like, and therefore become, a proton.

MATTER BIFURCATION

The Living Universe was no longer an antineutron; neither had it yet to become neutrons. It was a neutron-shaped *god particle* composed of an at-rest proton/antiproton pair. As soon as this particle was formed, the proton and antiproton aligned with each other and then split into a pair of photons. Because of the neutron shape, these two photons were intertwined in such a way that they combined and split into a pair of "god" particles before they could escape into space as photons. The protons and antiprotons within these particles then aligned with each other and divided into four "god" particles. At this stage in the Living Universe the ratio of matter to antimatter was exactly equal. The antiproton is the antiparticle to the proton.

NEUTRON ERA

This bifurcation process continued for perhaps 2^{256} cycles. It stopped when the mass of the antiprotons decreased to the point where they became electrons that can no longer annihilate with protons. On this last bifurcation cycle, the "god" particles all transformed into neutrons that were simply protons with electrons held inside them. The Living Universe had become a large expanding homogenous cloud of 2^{256} stable neutrons. As this neutron cloud expanded for billions of years, it became segmented by gravity into many individual clouds on many different scales.

THE TRANSFORMATION OF MASS AND ENERGY

As the electron particles lost mass within the neutrons, that mass did not disappear but rather was converted into increased rotational and vibrational energy. The mass of the neutron remained constant as its rest mass was converted into the kinetic mass of internal inertial motion.

THE FIRST POSSIBLE ATOMS

As time passed, the electron's mass got smaller and the neutron's internal energy grew larger and the proton/electron mass ratio got larger. When this ratio reached 146.5 to one, all of the neutrons in these vast clouds became radioactive and began to decay into protons and electrons. Soon the Living Universe became a seething mass of electrons and protons that were coupling together into hydrogen atoms and emitting photons as they dropped down into their ground states.

The reason that the neutrons became unstable at this particular point in their evolution has to do with the circlon shapes of the proton and electron. When the neutrons were first formed, the electron fit snugly inside the proton's secondary coils. As the electron continued to lose mass and increase in size the fit became tighter and tighter. Finally, when the ratio reached 146.5

to one, the electron became too big to remain trapped inside of the proton and popped out. For the first time in the history of the universe, it became possible for electrons to attach to the outside of protons and form atoms. The transition occurred when the mass of the proton, times the fine structure constant, divided by the mass of the electron, equals more than one: $M_p \alpha / M_e \geq 1$.

NUCLEAR SYNTHESIS

As this Grandfire occurred within the neutron clouds, the rapidly moving protons occasionally collided with neutrons and combined to form deuterons and tritons and alpha particles. These in turn combined with each other and with more neutrons to form the heavier elements. Once they were contained within nuclei, most of the neutrons became stable. By the time all of the free neutrons had decayed, the Living Universe was mostly clouds of hydrogen and helium with a sprinkling of the many stable isotopes of the heavier elements. Today there are 282 stable isotopes, but at this time in the evolution of the universe there may have been many more, since nuclear reactions required much less energy than they do today.

THE 2.7°K COSMIC BLACKBODY RADIATION

The rules of quantum electrodynamics show that with a proton/electron mass ratio of 146.5 to one, the photon spectrum of the hydrogen atom would have a temperature of 2.7°K. These 2.7°K blackbody photons filled the universe near its beginning and they still fill it today.

GALAXY FORMATION

The Living Universe took many more billions of years for the clouds of atoms to assemble into stars and galaxies. As they grew old, some of the larger stars ended their lives with type Ia supernova explosions.

DARK ENERGY

Today when we observe very old and far away supernova explosions, they are found to be much less energetic than the same type of supernovas that we might observe in neighboring galaxies. The reason that supernovas were less energetic in the distant past is because ancient atoms produced atomic spectra photons with much less energy and longer wavelengths than the same atoms do today. It is these weaker and weaker atomic energy levels that has been called "dark energy."

THE HUBBLE SHIFT

It is this effect that is the cause of the Hubble cosmological redshift. Photons from older and more distant galaxies have longer and longer wavelengths and smaller energies. We can calculate the rate of electron mass transformation by measuring the Hubble constant to greater and greater distances.

CONSERVATION OF MASS

In the evolution of the Living Universe, mass and energy were not created. They remained constant throughout the whole process. The universe began with a *single* anti-hydrogen atom that had a mass of 2^{256} protons. Today the mass is basically the same with the universe containing 2^{256} protons. All of the protons and electrons produced in the initial bifurcation cycle are all still with us today. The creation process was 100% efficient.

THE BIG BANG

This scenario for the Living Universe is compatible with all of the physical evidence that has been presented as support for the "standard" model of the Big Bang. It accounts for each component of the Big Bang theory but at the same time avoids the many paradoxes and contradictions inherent in Big Bang cosmology.

ACKNOWLEDGMENTS

When I set out to write this book, I didn't know what I was doing. Science writing generally didn't extend to taking "lunatics" seriously. How could I approach my subject with intellectual respect? In my search for literary role models, I kept coming back to two sources—Lawrence Weschler and Oliver Sacks—and I acknowledge them both for helping me to see how I could write outside the normative frame. Brian Rotman also has been an inspiration—his work has had an abiding effect on my life. The writer to whom I owe an incalculable debt is Augustus De Morgan, and his spirit permeates this book.

Very few people read this manuscript, but I thank them for giving me the one thing a writer needs above all else—honest feedback: my mother, Barbara Wertheim, my friends Tom Wagner and Dave Shulman, and my former husband, Cameron Allan, who was with me through most of my encounters with Jim.

My editor, Jackie Johnson, has been all that an editor can be—an avid reader, a perceptive critic, and a philosophical sounding board. It has been a joy to work with her. My agent, Michelle Tessler, took this book on as her first project when she started her own literary agency. She has waited a long time to see it come to fruition, and I hope it has been worth it. George Gibson was the only publisher willing to invest in this idea, and I thank him for his belief in both my subject and myself.

This book would not have been possible without the support of the Carter family. Their kindness and generosity throughout the long process of researching and writing have enabled me to do something that I many times thought would be impossible. To Jim, his sons, Paul and Eric, and above all to Linda, I offer my sincerest thanks.

My sister Christine lived through the process in all its manifestations, and it is hard to know how to thank her enough.

The film that I made with Cameron Allan, *It's Jim's World, We Just Live In It,* serves as a visual record of Jim's life and work during the critical period of late 1998 to early 2001, during which time he did his research on smoke rings. The documentary is a further resource for those interested in the subject and I thank Cameron for his beautiful photography and imaginative editorial ideas.

NOTES

Chapter 1: Under the Hood of the Universe

1. Jim's INCOBO engine was designed with the idea that it is always on, which means there is supposed to be no lag time waiting for water to heat up. The INCOBO was designed to run on diesel fuel, but Jim says it could easily be adapted to run on any kind of liquid fuel.
2. Ken Libbrecht maintains an extensive Web site about snowflakes and the physics of ice crystallization. It can be found at www.its.caltech.edu/~atomic/snowcrystals/. In addition to his research, Libbrecht has been at the forefront of developing techniques for photographing snowflakes. His books, including *The Snowflake: Winter's Secret Beauty*, contain gorgeous photos by himself and his colleague Patricia Rasmussen. On his Web site are beautiful time-lapse videos of different kinds of snowflakes growing in his ice crystallization chamber.
3. Geometry is currently going through a renaissance, in part because of the computer animation industry. When animators are modeling the drape of clothes and the movement of skin,

their software uses sophisticated geometrical techniques. There is now quite a demand in Hollywood for professionally trained geometers.

4. It is interesting to note that, as the Goodsteins discovered in the course of researching their book, James Clerk Maxwell also performed this proof and came up with the same version as Richard Feynman. It is not clear if Feynman knew about Maxwell's proof, but this is a thread that links three of the world's most brilliant physicists.

Chapter 2: Counterpart Universes "Excisting"

1. I worked as a science and technology columnist for *Follow Me* magazine from 1983 to 1986. To my knowledge, this is the first time any women's magazine has had a regular science columnist. I owe this remarkable fact to Robyn Powell, *Follow Me*'s enlightened editor, who believed, as do I, that women who read fashion magazines are also hungry for knowledge. Robyn interpreted her role in a dual sense: On the one hand, she had to provide her readers with entertainment, fashion tips, and fun; on the other hand, she saw that she had a unique opportunity to guide her audience into areas of understanding that they would not encounter elsewhere.
2. The bookstore was run by Michael Thompson and his wife, Kathleen. I am eternally grateful that Kathleen saw fit to pass Jim's package on to me.
3. The exhibition I curated of Jim's work at the Santa Monica Museum of Art, entitled "Lithium Legs and Apocalyptic Photons," was held during the summer of 2002. I thank the mu-

seum's director, Elsa Longhauser, for inviting me to do the show. As a curator, Longhauser was one of the first people in the United States to champion outsider art, and I am fairly sure she is the first museum director to devote an exhibition to outsider science.

4. Walter Murch's interest in Bode's law is allied with his interest in music. He has developed a hugely elaborate theory about how the planetary distances in our solar system, and the distances of the moons revolving around the planets, are analogs of musical scales. This aspect of Murch's work follows a long tradition of astronomers who have linked celestial motions to ideas about musical harmony, the most famous exponent of this concept being Johannes Kepler, the founder of modern astrophysics and an important precursor to Isaac Newton.
5. Dr. Simanek uses the ideas in the Museum of Unworkable Devices to test his students' understanding of the basic laws of physics. Can his students figure out what is wrong with these schemes? When Albert Einstein was working as an engineer at the Swiss Patent Office, one of his jobs was also to find the flaws in perpetual motion schemes sent in for patenting. Apparently he enjoyed the activity immensely. Simanek's wonderful museum may be found at www.lhup.edu/~dsimanek/museum/unwork.htm.
6. The French case involves two brothers, Igor and Grichka Bogdanov. In 1995, the Bogdanovs published several papers in peer-review physics journals that outlined a radical theory about the origin of the universe. After the papers were published, the veracity of their work was called into question and eventually discredited. Some physicists have claimed that the whole episode was a hoax, but the Bogdanovs have defended

their work as serious physics research. Today the brothers are widely known in France as science popularizers and as television show hosts.

Chapter 3: A BUDGET OF PARADOXES

1. Around 1875, colleges in Oxford and Cambridge abandoned the requirement for theological tests. Isaac Newton himself had been deeply troubled about taking such a test because he didn't accept the Trinitarian view of God. Newton was a Unitarian, which in the seventeenth century was regarded as a heretical position, and he considered dropping out of Cambridge rather than sign on to a belief he didn't hold. Fortunately for us all, he decided to stomach the test.
2. The integers are the "whole" numbers, which include zero and the negative whole numbers. Thus the set comprises . . . −3, −2, −1, 0, 1, 2, 3 . . . The integers are a subset of what are called the "real" numbers, which includes, along with the integers, the rational numbers (or fractions), and the irrational numbers, such as pi, which cannot be reduced to a fraction. It turns out there is also a two-dimensional version of the real numbers, called the "complex" numbers. This much larger set includes the "imaginary" numbers, which are those involving the square root of −1. De Morgan worked on the mathematics of the complex numbers, which he referred to by the charming name of "double algebra." De Morgan also helped to develop the system of four-dimensional "quaternions," which may be thought of as a kind of double-double algebra.
3. Fourier's work on heat flow introduced the idea that all mathematical functions could be seen as a product of sine waves.

This sounds arcane, but the Fourier transform is an exquisitely beautiful piece of mathematics that lies at the heart of modern signal processing as well as digital simulation of music. Every time you make a cell phone call or listen to a DVD, you are reaping the rewards of its mind-blowing power. Fourier's work on this subject is said to be the most quoted mathematical research of all time. Holograms are also physical manifestations of the Fourier transform.

Chapter 4: There's Diggers, and There's Everyone Else

1. It is now believed by scientists that the Port Orford meteorite is apocryphal and that Dr. Evans perpetrated a hoax.

Chapter 6: Circlon Science

1. Faraday and his wife lived their lives in a small apartment in the Royal Institution's attic. To their great sadness, they never had children, and lived an extremely modest life. In spite of his fame Faraday remained a humble man, quietly devoting his Sundays to an obscure Quaker-like branch of Christianity called the Sandemanians whose radical theology challenged its members to question the foundations of belief. A case has been made by several Faraday biographers that his views about fields were in part an outgrowth of his religious experience, particularly his experience of the Sandemanians' commitment to questioning dogma. Just as Newton had once seen the invisible hand of gravity as a manifestation of divine providence, so it seems Faraday found theological resonances

in the idea of fields. In both Newton and Faraday's work we see important examples of the ways in which unorthodox religious faith can sometimes inform science.

2. Jim's periodic table can be purchased from his Web site at www.circlon.com.

Chapter 8: Creating the World

1. Tait managed to catalog "the first seven orders of knottiness," comprising all knots up to seven "crossings" of a string. He'd imagined he would catalog all possible knots, but it soon became clear that the number of possibilities was "vaster than his interest." Others obligingly took up the project, and in 1888 Reverend Thomas Kirkman sent Tait a list of all knots up to the tenth order, or ten crossings of the string. Soon Kirkman sent in another list with knots of the eleventh order—more than fifteen hundred types. Today mathematicians have classified knots up to the sixteenth order, of which there are more than 1.4 million different types. It is known now that the number of distinct knot types is infinite, and the classification project continues.

   In the nineteenth century, Thomson hoped that vortex knots would help to explain a new discovery—the fact that all atomic elements have a built-in fingerprint of light, a spectroscopic signature that allows scientists to detect when an element is present. Sodium, for example, has two distinct yellow lines, while magnesium has a red line. Using these light fingerprints, scientists can examine any chemical compound and say what it is made of. Using spectroscopic analysis, astronomers can determine the composition of a star. Tait and

Thomson hoped that vortex atoms would explain how spectral lines arise. They imagined that the lines inherent in each element's fingerprint were produced when its vortex knot vibrated, rather like the tones produced by a bell. Just as different shapes of bells produce different sonic signatures, so Thomson speculated that different shapes of vortex knots might produce different light frequencies as the knot quivered in the ether. One of the many resonances between Thomson's theory and Jim's is that Jim also explains atomic spectra by the vibration of his circlon rings.

Chapter 9: Gravity and Levity

1. Scientists have not taken much notice of Wolfram's ideas, but apparently architects have been interested.

Chapter 11: Swimming Physicists

1. A famous defense of the uselessness of high-energy physics research occurred when physicist Robert Wilson was called to justify the multimillion-dollar Fermilab accelerator to the U.S. Congress Joint Committee on Atomic Energy in 1969. Wilson told Congress that the device would have no application whatsoever for national security; rather, he said: "It has only to do with the respect with which we regard one another, the dignity of men, our love of culture. It has to do with: Are we good painters, good sculptors, great poets? I mean all the things we really venerate in our country and are patriotic about. It has nothing to do directly with defending our country except to make it worth defending."

2. Twenty-six-dimensional versions of string theory are allied with a mathematical object called "the Monster" symmetry group, one of the more amazing structures in mathematics. Mathematician Mark Ronan has written a very fine book on the subject called *Symmetry and the Monster.* In 2010, I wrote an article summarizing the subject for *Cabinet* magazine, which is a short introduction to the idea of symmetry groups and their role in mathematics and physics.
3. I am indebted for this lovely allusion to a review of *The Hidden Reality* by Charles Seife published in the February–March 2011 edition of *Bookforum*. Seife himself is a wonderful writer about mathematical subjects and has written a book called *Proofiness* that explores the ways in which people misuse and abuse mathematical reasoning.

BIBLIOGRAPHY

In thinking about the subject of outsider physics, I cast about for precedents. Most of what has been written previously is cursory. By far the most useful resource I encountered is Augustus De Morgan's *A Budget of Paradoxes* from 1872, now available in Dover reprint. The other critical resource has been the works of outsider theorists themselves. I include here a full list of Jim Carter's books. In my own collection of theories I have around a hundred works, the majority of which are not publicly available. The most comprehensive source for dissident physics theories today is the Web site of the Natural Philosophy Alliance, www.worldnpa.org. As of 2011, the NPA's World Science Database contains a listing of more than nineteen hundred dissident theorists, a catalog of more than thirteen hundred books, and abstracts of more than five thousand papers. This database includes a separate Web page for each book and links to where it can be purchased.

Books by James Carter

Rogue River's Black Fortune (1965)
Gravity Does Not Exist (1970)

Gravitation Does Not Exist (1971)
The Cosmic Ring (1974)
The Circlon Atom (1975)
The Four Sexes (1988)
The Other Theory of Physics (first published 1993; updated every year, and sometimes every week, since then)
The Living Universe (2011)
Along a Twisted Trail of Truth: The Autobiography of a Science Crackpot (2011)

Jim Carter's work may be accessed on his Web sites: www.circlon.com and www.living-universe.com. These Web sites serve as a hub for his ideas and include many diagrams and animations of his theories. Jim's books and wall charts may be purchased on these sites.

General Bibliography

Agur, Elie. *Poems of Space and Time.* The Hague, Netherlands: Pace Safortim, 1997.

Bernstein, Jeremy. *Cranks, Quarks, and the Cosmos.* New York: Basic Books, 1993.

Carroll, Lewis. *Through the Looking-Glass, or What Alice Found There.* New York: Clarkson N. Potter Inc., 1972.

Carroll, Sean. *From Eternity to Here: The Quest for the Ultimate Theory of Time.* New York: Dutton, 2010.

Dawkins, Richard. *Unweaving the Rainbow: Science, Delusion and the Appetite for Wonder.* New York: Houghton Mifflin, 1998.

De Morgan, Augustus. *A Budget of Paradoxes.* New York: Dover Publications, 1954.

Einstein, Albert. *The World As I See It.* New York: Citadel Press, 1979.

———. *The Principle of Relativity: A Collection of Original Memoirs of the Special and General Theory of Relativity*. With H. A. Lorentz, Hermann Minkowski, and Hermann Weyl. New York: Dover Publications, 1952.

———. *Relativity: The Special and General Theory: A Clear Explanation That Anyone Can Understand*. New York: Crown Publishers, 1961.

Foucault, Michel. "The Masked Philosopher." An interview with Michel Foucault by Christian Delacampagne. *Le Monde*, April 6–7, 1980.

Goodstein, David L. and Judith R. *Feynman's Lost Lecture: The Motion of the Planets Around the Sun*. London: Jonathan Cape, 1996.

Greene, Brian. *The Elegant Universe: Superstrings, Hidden Dimensions and the Quest for Ultimate Reality*. New York: Vintage Press, 1999.

———. *The Fabric of Cosmos: Space, Time and the Texture of Reality*. New York: Alfred A. Knopf, 2004.

———. *The Hidden Reality: Parallel Universes and the Deep Laws of the Cosmos*. New York: Alfred A. Knopf, 2011.

Hawking, Stephen. *A Brief History of Time: From the Big Bang to Black Holes*. New York: Bantam Books, 1988.

Hirschfeld, Alan. *The Electric Life of Michael Faraday*. New York: Walker & Co., 2006.

Huizinga, Johan. *Homo Ludens: A Study of the Play Element in Culture*. Boston: Beacon Press, 1955.

Jarry, Alfred. *Exploits and Opinions of Dr. Faustroll, Pataphysician*. Translated and annotated by Simon Watson Taylor. Boston: Exact Exchange, 1996.

Kauffman, Stuart. *At Home in the Universe: The Search for Laws of*

Self-Organization and Complexity. Oxford: Oxford University Press, 1996.

Krauss, Lawrence M. *Hiding in the Mirror: The Mysterious Allure of Extra Dimensions, from Plato to String Theory and Beyond*. New York: Viking, 2005.

Lederman, Leon. *The God Particle: If the Universe Is the Answer, What Is the Question?* Boston: Houghton Mifflin, 1993.

Libbrecht, Kenneth. *The Snowflake: Winter's Secret Beauty*. With photographs by Patricia Rasmussen. Stillwater, MN: Voyageur Press, 2002.

Lindley, David. *Degrees Kelvin: A Tale of Genius, Invention and Tragedy*. Washington, D.C.: Joseph Henry Press, 2004.

Lisi, A. Garrett, and James Owen Weatherall. "A Geometry of Everything." *Scientific American* 303, no. 6 (December 2010).

McGurl, Mark. *The Program Era: Postwar Fiction and the Rise of Creative Writing*. Cambridge, MA: Harvard University Press, 2009.

Millis, Marc G. and Eric W. Davis (eds.). *Frontiers of Propulsion Science*. Reston, VA: American Institute of Aeronautics and Astronautics, 2009.

O'Brien, Flann. *The Third Policeman*. London: Dalkey Archive Press, 2002.

Ondaatjee, Michael. *The Conversations: Walter Murch and the Art of Film Editing*. New York: Alfred A. Knopf, 2004.

Rado, Steven. *Aethro-Dynamics and Electromagnetism: Part 1, Magnetism*. Los Angeles: 2000.

Randall, Lisa. *Warped Passages: Unraveling the Mysteries of the Universe's Hidden Dimensions*. New York: Ecco/HarperCollins, 2005.

Romanyshyn, Robert D. *Technology as Symptom and Dream.* New York: Routledge, 1989.

Ronan, Mark. *Symmetry and the Monster: One of the Greatest Quests in Mathematics.* Oxford: Oxford University Press, 2006.

Sacks, Oliver. *Uncle Tungsten.* New York: Alfred A. Knopf, 2001.

Seife, Charles. "Following the Thread: Brian Greene Reaffirms His Faith in String Theory." *Bookforum* (February–March 2011).

Silver, Daniel S. "Knot Theory's Odd Origins." *American Scientist* (March–April 2006): 158–165.

Sittampalam, Eugene. *Theory of Everything.* New York: Vantage Press, 1998.

Smolin, Lee. *The Trouble with Physics: The Rise of String Theory, the Fall of a Science, and What Comes Next.* New York: Houghton Mifflin, 2006.

Society for the Diffusion of Useful Knowledge. *The Penny Cyclopaedia.* London: Charles Knight, 1835.

Strathern, Paul. *The Quest for the Elements.* London: Penguin Books, 2001.

Susskind, Leonard. *The Cosmic Landscape: String Theory and the Illusion of Intelligent Design.* New York: Little, Brown & Co., 2005.

Tait, Peter Guthrie. *The Unseen Universe; or, Physical Speculations on a Future State.* With Stewart Balfour. Whitefish, MT: Kessenger Publishing (originally published by Macmillan & Co., 1875).

———. *Lectures on Some Advances in Physical Science.* Elibron Classics, 2006 (originally published by Macmillan & Co., 1876).

———. *Principles of Mechanics and Dynamics* (formerly titled *Treatise on Natural Philosophy*). With Sir William Thomson. New York: Dover Publications, 1962.

Thorne, Kip. *Black Holes and Time Warps: Einstein's Outrageous Legacy*. New York: W. W. Norton, 1994.

Tyndall, John. *Faraday as a Discoverer*. New York: Thomas Crowell Co., 1961.

Volk, Greg (ed.). *Proceedings of the Natural Philosophy Alliance: 17th Annual Conference of the NPA, 23–26 June 2010 at the California State University Long Beach*. Long Beach, CA: Natural Philosophy Alliance, 2010.

Wertheim, Margaret. "Where the Wild Things Are: An Interview with Ken Millett." *Cabinet* 20 (2005): 17–21.

———. "Hunting a Mathematical Snark." *Cabinet* 34 (2010): 53–56.

Weschler, Lawrence. *Mr. Wilson's Cabinet of Wonders*. New York: Vintage Books, 1995.

Winchester, Simon. *The Professor and the Madman: A Tale of Murder, Insanity, and the Making of the* Oxford English Dictionary. New York: Harper Perennial, 1998.

Woit, Peter. *Not Even Wrong: The Failure of String Theory and the Search for Unity in Physical Law*. New York: Perseus Books, 2006.

Wolfram, Stephen. *A New Kind of Science*. Champaign, IL: Wolfram Media, 2002.

INDEX

A NOTE ON THE AUTHOR

MARGARET WERTHEIM is a science writer with degrees in physics and mathematics. She has written for the *New York Times*, the *Los Angeles Times*, the *Guardian*, and many other publications. She is the author of *Pythagoras' Trousers*, a history of the relationship between physics and religion, and *The Pearly Gates of Cyberspace: A History of Space from Dante to the Internet*. In her work pioneering new methods of science communication, she founded the nonprofit Institute For Figuring, through which she and her sister Christine organized the Hyperbolic Crochet Coral Reef project, a touring exhibition at the intersection of science and art. Visit her Web site at www.physicsonthefringe.com.